FIRE AND EMERGENCY MEDICAL SERVICES ERGONOMICS

A Guide for Understanding and Implementing An Ergonomics Program in Your Department

U. S. Fire Administration
Federal Emergency Management Agency
16825 South Seton Avenue
Emmitsburg, MD 21727

March 1996

This Publication was produced under Contract EMW-94-C-4459 for the United States Fire Administration, Federal Emergency Management Agency. Any information, findings, conclusions, or recommendations expressed in this publication do not necessarily reflect the views of the Federal Emergency Management Agency or the United States Fire Administration.

TABLE OF CONTENTS

Glossary

APPENDIX A - Annotated Bibliography

APPENDIX B - Sources of Information

APPENDIX C - Sample Ergonomics Program

APPENDIX D - Sample Fitness Evaluation Protocol

APPENDIX E - Worker Compensation Rules

APPENDIX F - Suggested Injury Reporting System

APPENDIX G - Example Cost Benefit Analysis

PREFACE

The application of ergonomics is increasingly gaining attention within U.S. industry as a means for improving the workplace and preventing chronic and sometimes debilitating medical disorders. Often the concept of ergonomics seems alien to the fire and emergency medical services because unlike most industrial work environments, the nature of the emergency response scene is forever changing and demands a less than routine approach to accomplish the job. Despite this thinking, the proper application of ergonomics to the fire and emergency medical services can have a very significant impact in reducing many of the injuries that occur on or off the fire ground or emergency medical scene. Consider that *approximately 40% of all fire fighter injuries over the period of 1990-1994 are classified as sprains, strains, and other musculoskeletal ailments,* many of which may be attributed to the poor design of equipment and other ergonomic concerns. These injuries often result in significant amount of loss time and other large costs to fire and EMS departments throughout the country. A large number of these injuries are preventable by implementing an ergonomics program as part of the department's overall occupational safety and health program.

Ergonomics should not be considered just another program to implement, especially with the shrinking available resources available to many departments. Instead, as this guide points out, *the ergonomic program should be used to improve existing approaches to fire fighter and EMT safety, tactics, equipment selection, personnel fitness, and injury reporting procedures.* The investment in time and resources for applying ergonomics as recommended in this guide can yield enormous benefits in terms of improved department morale, better response efficiency, and lowered medical costs.

Implementation of an ergonomics program should be given priority within each department for a variety of reasons. Not only will a cost/benefit analysis usually easily justify its costs, but the institution of ergonomic policies may become regulated with the introduction of a new ergonomics standards which is currently in draft form from the Occupational Safety and Health Administration (OSHA) or some other future workplace standard.[1]

It is likely that consideration of ergonomics within a department can take several levels of involvement from highly sophisticated programs to simple, common-sense approaches. *This guide is intended to provide the information that lets each department decide which approach is best for them given their existing ergonomic problems and available resources.* Nevertheless, certain elements of the program must be common to all departments. These include placing attention to risk assessment, equipment selection, personnel fitness, injury reporting, and follow-up to injuries. These elements mirror those recommended in the OSHA draft standard.

---

[1]At the time the final draft of this guide was prepared (March 1996), the adoption of an OSHA-based ergonomic standard was considered unlikely for next several years. However, references are made to the OSHA draft proposed ergonomics standard since it provides useful guidelines which can be applied to the fire and emergency medical services.

ACKNOWLEDGEMENTS

The authors of this manual would like to extend their appreciation to a number of individuals and organizations which contributed to the development of this guide. Several departments were contacted during an ergonomics survey and provided useful findings:

1. The Charlotte (North Carolina) Fire Department with specific input from Robert Tutterow, Safety and Logistics Officer;

2. The Fairfax County (Virginia) Fire Department with specific input from Captain David Carpenter, Safety Officer;

3. The Fire Department of New York City with specific input from Chief Al Santora and Lieutenant Gerard Speer, both of the Division of Safety;

4. The Seattle (Washington) Fire Department with specific input from Chief Rodney Jones (Chief of Safety);

5. The St. Louis (Missouri) Fire Department with specific input from Captain Brian Walsh, Fire Department Medical Officer;

6. The St. Paul (Minnesota) Fire Department with specific input from Terry Holter, of the Loss Control Department;

7. The Tulsa (Oklahoma) Fire Department will specific input from Mike Mallory, Department Safety Officer; and

8. The Virginia Beach (Virginia) Fire Department with specific input from Captain Murrey Loflin, Safety Officer.

In addition, two other individuals and their respective organizations provided advice and suggestions for this guide's development including:

- Richard M. Duffy, Director, Occupational Health and Safety Department, International Association of Fire Fighters (IAFF) who offered overall guidance for the development of fire/EMS department ergonomics programs; and

- Paul Crawford, formerly of the Riverside (California) Fire Department, who provided an example cost benefit analysis which served as a model for this guide.

Lastly, appreciation is extended to the City of Phoenix Fire Department whose programs and practices served as a model for much of the guide's development. Recognition is especially extended to Assistant Chief Steve Storment, Deputy Chief Van Somers, Deputy Chief Gary Morris, Deputy Chief Mike Ingallina, Division Chief Pat Baliecki, Battalion Chief Tim Gallagher, and Captain Scott Pelton.

This guide is intended to offer both small and large, career and volunteer departments, specific recommendations and examples for applying ergonomics. The guide's contents include:

- Chapter 1 provides an **introduction to ergonomics** in general and how it can be considered in the fire and emergency medical services;

- Chapter 2 lists a number of **medical disorders commonly associated with ergonomic hazards;**

- Chapter 3 gives the **basic steps for establishing a stand-alone ergonomics program** or implementing it as part of an on-going department safety program (Appendix C is a copy of a sample ergonomics program);

- Chapter 4 provides guidance on **how to identify ergonomic hazards** within your department;

- Chapter 5 offers **guidelines for preventing and controlling ergonomic hazards,** addressing a number of specific areas relevant to the fire and emergency medical services;

- Chapter 6 discusses the **role of training in an ergonomics program** and provides an outline of training information;

- Chapter 7 describes **aspects of a medical management system** including both preventative measures and post-injury treatment protocols (Appendix D provides a recommended protocol for fitness evaluations, while Appendix E covers rules on workers' compensation);

- Chapter 8 specifically deals with **methods for reporting injuries** and evaluating injury data (Appendix F provides a suggested injury reporting system);

- Chapter 9 gives **recommendations for how to specific implement an ergonomics program** in your department;

- Chapter 10 furnishes **techniques to evaluate the effectiveness of the ergonomics program** (Appendix G provides a method for conducting a cost-benefit analysis in your department),

The guide is also supplemented with a glossary of terms, an annotated bibliography of useful references (Appendix A) and a list of other sources of information (Appendix B).

Today's fire fighter faces an ever growing number of hazards. Many of these hazards are obvious as they have always been in fire fighting since its inception--the direct contact with fire, the excessive amounts of heat, and enormous physical burden that fire fighters accept in the course of their duties. Fire fighters face a variety of emergency situations, nearly always involving circumstances not of their choosing. Nevertheless, each year a significant number of injuries occur which are not from burns or exposure to other traditional fire ground hazards.

The U. S. Fire Administration (USFA) maintains statistics on fire fighter injuries over the past several years through a national voluntary reporting system. Surprisingly, only 20% of fire fighter injuries are from burns and smoke inhalation (Figure 1-1 provides statistics over period 1990 to 1994). *The leading number of fire fighter and emergency medical technician (EMT) injuries (39.7%) are due to strains, sprains, and muscular pain from overexertion or falls.* The vast majority of these injuries affect the responder's torso (trunk) and feet. These injuries are a significant proportion of all injuries which occur during activities on the fire ground (35.9 %) and non-fire emergencies (48.7 %), but even a greater proportion for emergency responders when responding to and returning from an incident (53.6%) as shown in Table 1-1.

This pattern of injuries conclusively shows that fire fighters are more likely to receive sprains, strains, and other muscular pain injuries than other types of injuries. Certainly fire fighters must carry heavy equipment and be well protected; however, less attention may be given to aspects of the work environment not traditionally thought to be serious hazards. The high incidence of strains and sprains should be an indication that the application of ergonomics is strongly needed in the fire and emergency medical services to avoid more of these injuries.

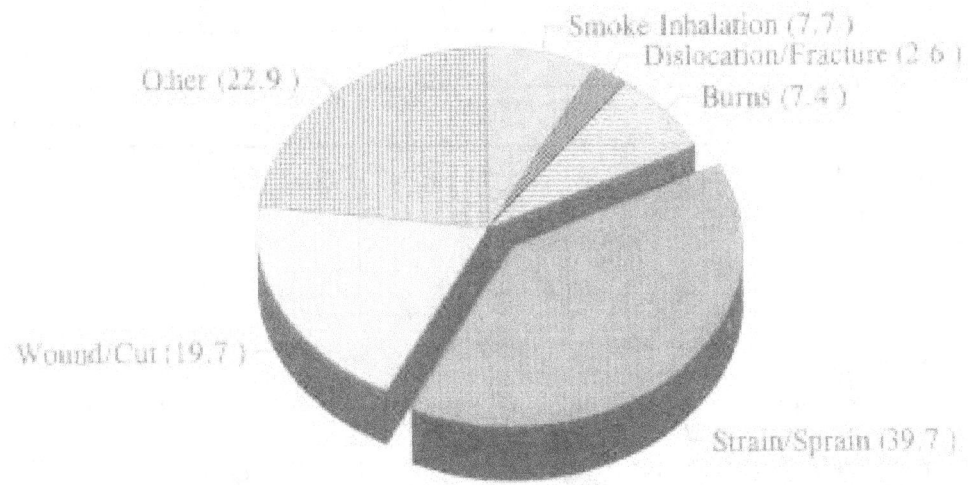

Figure 1-1. Types of Fire Fighter Injuries

SOURCE: 1989-1994 NFIRS Injury Reports (based on 371,950 fire fighter injuries).
NOTE: Includes only injuries that occurred responding to/returning from, on the fire ground, and during non-fire emergencies.

Table 1-1. Fire Fighter Injuries: Type of Injury by Type of Activity

Type of Injury	Type of Activity			
	Responding To/ Returning From Incident	On Fireground	During Non-Fire Emergency	Other†
Burns (fire or chemical)	0.7%	9.9%	0.7%	3.3%
Asphyxiation (inc. smoke inhalation	1.7	9.8	2.3	1.0
Wound, cut, bleeding, bruise	23.6	19.9	17.0	22.6
Dislocation, fracture	5.7	2.4	2.1	3.7
Strain, sprain, muscular pain	53.6	35.9	48.7	49.9
Other*	14.7	22.1	29.2	19.5
Total	100.0	100.0	100.0	100.0
Number of cases with known type of activity and type of injury	28,575	270,980	72,395	126,250

† Includes training plus "other on-duty" injuries

* Includes respiratory distress, eye irritation, heart attack or stroke, and thermal stress

SOURCE: 1990-1994 NFIRS data (based on 371,950 fire fighter injuries with activity reported.

Defining Ergonomics and Developing Awareness

Ergonomics is a developing science which is aimed at enhancing the healthy interrelationship of humans, machines, and the work environment. Simply put, *ergonomics is the practice of making the work environment safer and more productive for the worker.* The first step in developing any ergonomics program is to create among workers, at all levels in a given field of endeavor, the ability to recognize ergonomic problem areas within their own profession. The ability to recognize potential problem areas is followed by learning what actions should be taken to correct the problem. Such an ergonomic program is said to be *proactive.* That is, through the application of what are sound preventive actions, the injury rate or the psychological feeling of well-being associated with a given job is expected to be improved. The ergonomically-conscious fire fighter or EMT and their supervisors must learn to look for several factors when performing an analysis of (human) risks associated with a job.

GENERAL WORKPLACE ERGONOMIC RISKS

1. **forceful exertions, e.g.,**
   - lower back - lifting heavy weights away from the body
     arm/hand - pinch grip, forced or prolonged and repeated deviation at wrist

2. **awkward postures, e.g.,**
   - spinal rotation under power
   - shoulder elevations under power
   - prolonged seated work
   - elbows elevated
   - deviated wrists

3. **localized contact stress, e.g.,**
   - body part contacts unpadded sharp edge
   - chair seat pan cuts into thigh at popliteal area
   - grasping small diameter tools
   - using body part as striking tool

4. **vibration, e.g.,**
   - extended operation of power hand tools
   - heavy equipment operation, usually seated

5. **environmental conditions, e.g.,**
   - heat - heat stress/hyperthermia
   - cold - cold stress/hypothermia, reduced dexterity and tactility, frostbite
   - particulate agents - smoke, dust, snow, sleet, etc.
   - noise
   - liquid and gaseous agents

6. **repetitive/prolonged activity,** e.g., tasks with short cycle times

---

Awareness must be supported by analysis techniques for more subtle problem areas. In specific jobs, complex analyses may not be needed, since the history of on-the-job injuries will make obvious the need for changes. Programs which deal primarily with known problem jobs are said to be *reactive*. In order to accomplish efficiency in awareness, *both management and labor must be trained to analyze the tasks involved in performing their jobs and thereby to reduce the incidence of instantaneous trauma and long term, repetitive strain injuries on-the-job*.

Brief History of Ergonomics

The Ergonomics Society, which was formed in England around 1960, was over twenty years old before advertisers discovered the term and applied it to various consumer items in this country. In this country ergonomics has strong ties with human factors, human factors engineering and work physiology. The U.S. Human Factors and Ergonomics Society is more oriented toward the psychological aspects of the relations among humans, machines and the environment than is the European equivalent. Meanwhile, international societies, *e.g.*, International Foundation for Industrial Ergonomics and Safety Research, and several new scientific journals, e.g., *Applied Ergonomics, International Journal of Industrial Ergonomics,* etc., have appeared in recent years. As in any other relatively young discipline, its practitioners, the so-called ergonomists, are still exploring the limits of a profession.

The original impetus which ultimately gave birth to ergonomics was the drive to improve the effectiveness and the efficiency of equipment used in warfare, specifically during World War II. So strong were its ties to the military in this country that it has been only recently that American industry has come to recognize that the application of ergonomic principles could also reap benefits for them. Industry is learning that through that through the redesign of jobs and/or workstations, better products may be produced, efficiency and wellness of the worker also improves. The bottom-line for management is the cost in days lost on-the-job plus the high cost of hospitalization or even out-patient treatment. For the worker, it is a more pleasant work environment plus reduced risk of short or long term injury.

Relationship of Ergonomics to Fire Fighting and EMS

Of course, there are considerable differences between the tasks performed by fire fighters and emergency medical service (EMS) personnel and workers in most other occupations. But injury still results in lost time and/or poor performance on-the-job. In addition, ergonomic factors can have just as much or more impact on the operation of a fire department than they do in any other industry. It is true that often fire fighter and EMS injuries occur in a workplace that is dynamic, environmentally stressful and fraught with high risk. In many of these conditions, ergonomics can play little, if any, role in preventing injury. Nevertheless, *it is through learning to recognize poorly designed equipment and unsafe job practices that improvements can be made, and injury and lost time be lessened.* This is one profession where, if ergonomics can prevent a few on-the-job injuries, the ramifications may have immediate impact on the life safety and well-being of the public at large. It is toward such a goal that an ergonomic program must be directed.

A number of fire fighter activities are easily targeted for an analysis of ergonomic hazards. For example:

- Fire fighters are frequently injured when **lifting or carrying** fire hose, ladders, or victims. Some of this activity occurs outside the fire ground and may not involve emergency conditions.

- By the nature of the hazards, fire fighters and EMTs must assume **awkward positions,** particularly in fire suppression as they keep low to the ground to avoid rising heat and the hotter temperatures in a room or in having to remove victims from cramped spaces cluttered with debris. Fire fighters must also face enormous heat burns and risk the potential for heat stress. These same conditions should be avoided before and after the emergency activity by controlling the activities or conditions where they can be controlled.

- A significant number of injuries occur related to **vehicular** use. Step heights must be considered as well as the position that the fire fighter must ride when going to or from the emergency scene.

- Emergency medical workers often have to provide emergency patient care at the site of the emergency, often involving **improper work surfaces** or **difficult environmental conditions.**

- As in other industries, the fire and emergency medical services are increasingly dependent on video display terminals (VDTs). These devices are well known to produce eye strain which can lead to vision impairment if left unattended. **Poor VDT workstation design** and **continuous long** use in a bad posture leads to muscular disorders.

- Fire fighter protective clothing and equipment can reduce mobility, hand dexterity, and tactility while producing conditions which may lead to **heat stress. Improperly designed tools** which must be carried to and from the scene can be difficult to transport or use. More attention must be placed on the design and correct ergonomic use of these items.

The literature survey and preparation of the annotated bibliography for this guide were extremely important to identify key references and information which could be applied to the fire and emergency medical services. A review of this literature revealed the following:

1. ***Principal ergonomic concerns relevant to fire fighting and emergency medical services include injuries associated with strains and sprains.*** This category has consistently remained the most often reported injury on fire service statistics compiled by the International Association of Fire Fighters (IAFF), National Fire Protection Association (NFPA), and U.S. Fire Administration (USFA) over the past several years. While these injuries are not always as severe as other types, they are the number one cause of lost time.

2. ***Common causes of ergonomic injuries and work disorders include heavy lifting (hoses, ladders, and victims) and awkward positions (such as stepping out of fire apparatus).*** In many cases, these injuries may have been preventable with proper fitness and training, but many circumstances or the lack of proper equipment also contribute to ergonomic injuries. The fire service is also subject to many of the other hazards common in related industries such as excessive noise, heat stress, and vibration.

3. ***Related concerns and similarities of concerns, injuries, and work disorders have been described for related industries and work professions.*** The most significant difference between the fire/emergency medical services and other industries is the emergency nature of the job. Whereas the work environment in other industries is easier to control because it is precisely known, the fire fighter or other first responder cannot always choose where they must perform their duties.

4. ***Few references reported specific incidents where lack of appropriate ergonomic controls or practices contributed to an injury.*** However, there were several generalizations which provided descriptions of practices which often result in injury. Often authors provided guidelines for the avoidance of injuries in these areas primarily through workplace design and proper design/selection of equipment and tools.

5. ***There were a number of references which described types of products or practices in the market place and how their design reduced ergonomic work disorders.*** Some of these articles bordered on the commercial side, but most were informative in explaining how properly design equipment or work practices could reduce ergonomic hazards. Most of the information in this area was criticism of poorly designed tools or practices such as the "one-size-fits-all" concept. In other cases, there was debate on the utility of certain practices. For example, opinion is split whether back supports provide any real help in reducing back injuries.

6. ***Only a few of the articles or references provided standards or guidelines for clothing/equipment selection as related to ergonomics.*** These sources were generally very specialized and a few applied to the fire services primarily for the selection/design of protective clothing and fire apparatus. Nevertheless, there were a number of articles prepared on the issue of back injuries and heat stress in the fire service.

7. ***Most existing ergonomic programs are in place for manufacturing industry, some in anticipation of possible OSHA regulations on ergonomics.*** A number of factors seem common for the effectiveness of these programs, including employee education, management commitment, adequate injury reporting systems, a process to initially identify ergonomics hazards, and review procedures to follow up on recommended controls.

8. *General Practices in use by fire departments as reported in the literature related to ergonomic concerns, including mainly awareness campaigns and recommendations for proper fire fighter or EMT fitness levels.* Most departments consider ergonomics as part of the overall occupation safet and health program but do not have specific ergonomics programs in place.

9. *While most departments have injury reporting systems, few utilize this information properly for identifying and correcting hazards.* A review of the injury reporting systems for several departments showed difficulty in extracting data which would permit determination of problem jobs or activities.

Additional specific interpretations of the identified literature have been used to assist in the analysis of injury data, evaluation of ergonomic practices, and interpretation of industry standards for supporting the development of this guide.

The annotated bibliography for this guide in **Appendix A** provides several references to articles and books which may be useful to department which want to obtain additional details beyond this guide. The annotated bibliography is divided into subject areas and a short description accompanies each citation. Other useful sources of information are provided in **Appendix B.**

CHAPTER 2 - ERGONOMIC-RELATED DISORDERS

Most of the industry focus for ergonomic-related disorders is on cumulative trauma disorders (CTDs) which typically affect nerves and tendons in upper extremities. In the fire service, while CTDs are encountered, most often back pain, strains, and sprains are the result of injuries, many of which are produced by acute event (e.g., falling, slipping, overexertion during lifting). Other forms of ergonomic-related disorders include heat stress and noise stress. Since many fire and emergency medical service administrative functions involved office environments, concerns for the ergonomic hazards of video display terminals exist as well.

Information is presented in this chapter for describing the more common ergonomic-related disorders and their causes or risk factors which affect fire fighters and EMTs.

Cumulative Trauma Disorders (CTDs)

Cumulative trauma disorders or CTDs can be described as wear and tear on joints and surround tissue because of overuse. Every joint in the body potentially be affected but the lower back and upper limbs are the areas that receive the most injuries. Cumulative trauma is in contrast to acute trauma as may occur from cuts and falls. Cumulative disorders accumulate through time, hence the term.

Cumulative trauma is also referred to by a variety of terms, such as *musculloskeletal disorders, overuse syndrome, or repetitive motion disorders.* The terms are not completely synonymous, in that disorders such as hearing loss, eyestrain, and joint diseases like arthritis can be included.

These disorders are sometimes called illnesses. This terminology originates from the OSHA 200 Log form which classifies medical problems as either injuries or illnesses. Acute problems are classified as *injuries* and chronic problems as *illnesses.* In this framework, CTDs fall into the illness category and are consequently are sometimes called illnesses.

Most often, cumulative disorders are mild and temporary. But in their more severe forms, CTDs can be very painful and sometimes permanently disabling. The disorders can interfere with all aspects of daily life, sometimes to the point where the simplest of tasks become unmanageable.

Cumulative trauma disorders affect any area of the body where tendons, joints, and nerves are found. Most commonly, CTDs target the upper extremities which include all of the anatomical components from the shoulder to the fingers. *While acute injuries resulting from a single event do occur to the upper extremity, more disorders are being recognized as the cumulative effect of multiple small, often unrecognized, repetitive injuries, particularly those for the back.* Upper extremity CTDs largely affects the origin of muscles (where the muscle attaches to the bone, the tendon, the joints, the blood vessels, and the nerves).

Types and Symptoms of CTDs. There are several types of CTDs. The tables on the following two pages list several common disorders. Most CTDs fall into two main categories:

1. **Tendinitis.** Tendons serve as links that connect muscle to bone and come into play whenever a muscle is used for motion of a bone structure. In some areas of the body, tendons slide through sheaths. As with any other moving part, overuse of tendons can cause friction which cause wear and tear, and expansion or swelling. When tendons or their sheaths swell, there is pain and tenderness, known as tendinitis.

2. **Nerve Compression. Nerves are** found throughout the body and several points exist where it is possible for nerves to be compressed. Pinching of nerves is often caused by making certain awkward motions or assuming certain postures. Other times, compression can be caused by swelling of nearby tendons.

The overall occupational prevalence of upper extremity CTDs is unknown, but data collected for individual work sites, medical clinics, and insurers show a significant increase in these types of cases. Because the disorder usually develops slowly and with minimal symptoms, determining the true incident rate for occupational related disorders is difficult. Largely, the ambiguity exists because causality is hard to confirm. This is equated to the latency period, which is the time between exposure and the onset of disease. Nevertheless, unlike many occupational diseases with latency periods, the exposure in CTDs continues. A further complication is that most of these disorders are preceded by many vague symptoms that are not commonly thought to be precursors to the disease. In addition, some physicians may not yet be familiar with the relationship between these disorders and the work environment, especially the rigors for fire fighting and emergency medical response.

Although there are differences based on the exact type of disorder and the part of the body affected, the symptoms of CTDs in general include:

- soreness, pain, and discomfort;
- limited range of motion;
- stiffness in joints;
- numbing, tingling sensations ('pins and needles');
- popping and cracking noises in the joints;
- 'burning' sensations;
- redness and swelling; and
- weakness and clumsiness.

CTDs are very often presented as fatigue or other minor discomfort that are relieved by rest, It is also common for these conditions to be presented as minor injuries or complaints that are presumed not to have long term consequences. Since recognizing CTDs in these early phases is difficult, the potential for both over-reporting and under-reporting exists.

SYNOPSIS OF CUMULATIVE TRAUMA DISORDERS

Hand and Wrist

- **Tendinitis** - Inflammation of a tendon.

- **Synovitis** - Inflammation of a tendon sheath.

- **Trigger finger** - Tendinitis of the finger, typically locking the tendon in its sheath causing a snapping, jerking movement.

- **DeQuervain's disease** - Tendinitis of the thumb, typically affecting the base of the thumb.

- **Ganglion cyst** - Synovitis of tendons of the back of the hand causing a bump under the skin.

- **Digital neuritis** - Inflammation of the nerves in the fingers caused by repeated contact or continuous pressure.

- **Carpal tunnel syndrome** - Compression of the median nerve as it passes through the carpal tunnel.

Elbow and Shoulder

- **Epicondylitis** ('tennis elbow') - Tendinitis of the elbow.

- **Bursitis** - Inflammation of the bursa (small pockets of fluid in the shoulder and elbow that help tendons glide).

- **Rotator cuff tendinitis** - Tendinitis in the shoulder.

Neck and Back

- **Tension neck syndrome** - Neck loreness, mostly related to static loading or tenseness of neck muscles.

- **Posture strain** - Chronic stretching or overuse of neck muscles or related soft tissue.

- **Degenerative disc disease** - Chronic degeneration, narrowing, and hardening of a spinal disc, typically with cracking of the disc surface

- **Herniated disk** - Rupturing or bulging out of a spinal disc.

- **Mechanical back syndrome** - Degeneration of the spinal facet joints (part of the vertebrae).

- **Ligament sprain** - Tearing or strength of a ligament (the fibrous connective tissue that helps support bones).

- **Muscle strain** - Overstretching or overuse of a muscle.

- **Radial tunnel syndrome** - Compression of the radial nerve in the forearm.

- **Thoracic outlet syndrome** - Compression of nerves and blood vessels under the collar bone.

Legs

- **Subpatellar bursitis** ("housemaid's knee" or "clergyman's knee") - Inflammation of patellar bursa.

- **Patellar synovitis** ("water on the knee") - Inflammation of the synovial tissues deep in the knee joint.

- **Phlebitis** - Varicose veins and related blood bessel disorders (from constant standing).

- **Shin splints** - Microtears and inflammation of muscle away from the shin bone.

- **Plantar fascitis** - Inflammation of fascia (thick connective tissue) in the arch of the foot.

- **Trochanteric bursitis** - Inflammation of the bursa at the hip (from constant standing or bearing heavy weights).

Risk Factors. There are several factors that can increase the risk of developing CTDs. The more factors involved and the greater the exposure to each, the higher the chance of developing a disorder. These factors include both working conditions and personal issues as listed below:

Working Conditions:

1. *Repetition* - risk increases with number and frequency of motions made by a particular part of the body.

2. *Force* - risk increases with the amount of exertion required for particular motions.

3. *Awkward postures* - risk increases with positions of the body which deviate from a neutral position; primarily bent wrists, elbows away from their normal positions at the side of the body, and a bent or twisted lower back.

4. *Contact stress* - risk increases with excessive contact between sensitive body tissue and sharp edges or unforgiving surfaces on a tool or piece of equipment.

5. *Vibration* - risk increases with exposure to vibrating tools or equipment, whether a hand-held power tool or whole-body vibration.

6. *Temperature extremes* - risk increases with exposure to excessive heat or cold.

7.   *Stressful conditions* - risk increases with certain stressful situations at work or due to the nature of the work.

Personal Issues:

1.   *Off-the-job activities* - activities at home or in leisure can contribute to CTDs.

2.   *Physical condition* - poor personal fitness can play a role in the development of some CTDs.

3.   *Other diseases* - factors such as gout and diabetes mellitus can also be implicated for CTDs occurring.

The supporting evidence that links these factors to CTDs is well documented in the scientific literature. Currently, the American National Standards Institute (ANSI) Z-365 Committee on Cumulative Trauma has compiled probably the best review of studies and data. This committee has reached consensus concerning these risk factors and has adopted a framework for these factors which is equivalent to that presented above.

The levels of exposure, i.e., how many repetitions, at what force, and in what posture, that can trigger a disorder are not yet known. Moreover, precision measurement of these factors is difficult. However, more is worse, and combinations of several factors increase the likelihood of disorders. Conversely, if risk factors can be reduced, the risk of a problem can be minimized.

Methods of Diagnosis. Obtaining a careful medical history is the most important step in diagnosis, The medical history should include questions about both workplace and personal factors. Leaving the impression that CTDs come only from activities associated with work is wrong, Clear examples of recreational injuries are: tennis elbow, golfer's elbow, and a variety of hand and wrist disorders that have been casually linked to excessive use of home computers or video games.

Details of diagnosis are beyond the scope of this guide; however, some appreciation of the principles applied may be helpful. The physical examination should been refined to suggest the appropriate diagnosis. Examples include palpation and determination of point tenderness. Additional maneuvers and tests which can be useful in diagnosis include:

Tinel's Test: Tapping over the median nerve at the wrist reproduces the numbness and tingling in the median nerve distribution suggesting Carpal Tunnel Syndrome.

Finkelstein's Test: Ulnar deviation of the hand is combined with the thumb flexed across the palm and the fingers flexed over the thumb. Severe pain results at the radio styloid due to stretching of the thumb abductors and flexor muscles, suggesting Carpal Tunnel Syndrome.

Phalen's Test; The patient is asked to put the backs of the hands together, causing the wrists to flex to near 90 degrees. If the patient has symptoms of median nerve problems (a numbness and tingling in the thumb through third fingers) within one minute, Carpal Tunnel is suspected. There is a reverse Phalen's with hyperextension.

Allen's Test: The patient is asked to make a tight fist. While the fist is clenched, the physician occludes both the radial and ulnar arteries of the wrist. The hand will become white when reopened. One vessel is released; if the hand turns pink, it can be assumed that a vessel is open. The process is repeated to check the other vessel. If the hand does not resume its normal pink color when one vessel is released, its suggests that the vessel is occluded. Occlusion of vessels is common, when the heal of the hand is repeatedly used as a pounding tool.

Adson's Test: This maneuver is sometimes helpful to evaluate Thoracic Outlet Syndrome. The radial pulse is found at the base of the thumb. The patient is then asked to turn the head slowly, but forcibly, to the opposite side. If the pulse is lost, narrowing of the outlet can be suspected.

Exaggerated Attention: This is also used to evaluate the Thoracic Outlet Syndrome. With the pulse monitored, the patient draws the shoulders back to exaggerate the "at attention." If the pulse disappears, the diagnoses can be suspected.

Nerve Conduction Studies: This is not to be confused with electromyography. Nerve conduction studies measure the speed at which an electrical impulse is transmitted by the nerve. A nerve that is functioning less than normally, for whatever reason, will be apparent. This diagnostic tool is commonly used to identify Carpal Tunnel Syndrome.

Vibrometry: This technology has been an experimental model for many years. The postulate is that when peripheral nerves (such as the median nerve) are compromised, the ability to perceive vibration is one of the first functions to fail. Thus, the patient can demonstrate measurable changes in vibratory sensation, detecting impending damage to the nerve may be possible. This may permit earlier intervention and avoid serious nerve damage.

Back Disorders

The back is a complex system consisting of several distinct spinal regions. Lifting, bending, and twisting motions (on or off the job) can cause severe injury and pain. Because the lumbar region is at the greatest risk during normal work, it deserves to be the main focus of back conservation and maintenance. Next to the common cold, back disorder is the reason most often cited for job absenteeism. According to the 1994 International Association of Fire Fighters' Death and Injury Survey, sprains and strains, of which back injuries comprise a major share, were the leading cause of line-of-due injuries (49.7%). Back injury was also the largest contributor to disability retirements resulting from line-of-duty injuries (53.2%). This highlights the fact that these injuries are chronic and account for a significant loss time and large medical expenses.

Types of Disorders. Pulled or strained muscles, ligaments, tendons, and discs are perhaps the most common back problems. They may occur to almost half the work force at least once during a lifetime. Most back disorders result from chronic, or long-term, injury rather than from one specific incident. When back muscles or ligaments are injured from repetitive pulling and straining, the back muscles, discs, and ligaments can become scarred and weakened and lose their ability to support the back. This condition makes additional injuries more likely.

A synopsis of the conditions which lead to back injuries are listed below.

SYNOPSIS OF CONDITIONS RESULTING IN LOW BACK PAIN

- **Lumbosacral strain** - caused by overuse of the muscles of the lumbar and sacral areas of the back.

- **Lumbosacral sprain** - caused by overuse of ligaments in the lumbar and sacral areas of the back.

- **Postural low back pain** - results from overuse of the lumbosacral muscles by maintaining a posture that requires these muscles to work beyond their capabilities.

- **Muscular insufficiency** - occurs when muscles are unable to bear stresses required of them.

- **Sacroilitis** - caused by inflammation of the joints between the lowest spinal bones (sacrum) and the hip bones (ilium).

- **Herniated disc** - results when the disc that sits between two spinal bones (vertebrae) bulges from between them.

- **Degenerated disc** - results when wear and tear on disc slowly destroys its structure.

Risk Factors. Back disorders are frequently caused by the cumulative effects of faulty body mechanics such as: excessive twisting, bending, reaching, lifting loads that are too heavy or too big, too far from the body, staying in one position for too long, poor physical conditions, and poor posture. Prolonged sitting stresses the body, particularly the lower back and the thighs. It may cause the lower back to bow outward if there is inadequate support. This abnormal curvature can lead to painful lower back problems, a common complaint among office workers. Several factors may be related to back disorders which include both those which are related to the workplace or job or to personal attributes.[2] Included in this list are:

---

[2]Arun Garg and J. Steven Moore, "Epidemiology of Low-Back Pain in Industry," Occupational Medicine: Stat of the Art Reviews, Vol. 7(4), October-December 1992, Philadelphia, Hanley & Belfus, pp 593-608.

Job Risk Factors

1. *Heavy physical* work - Information based on workers' compensation claims and insurance data show that low-back pain is more prevalent in heavy physical jobs; the potential for overexertion injuries is greater.

2. *Lifting* - Low-back pain is clearly triggered by lifting; the weight, speed, duration, and frequency of lifting affect the onset of low- back pain.

3. *Bending, stretching, and reaching* - Bending in combination with lifting appears to be the most common cause for low-back pain; the incident of low-back pain also increases with loads placed away from the body.

4. *Twisting* - Lifting in combination with twisting has been implicated in low-back pain injuries.

5. *Pushing and pulling* - Pulling and pushing account for 9 to 18% of all back strains and sprains.

6. *Prolonged sitting and standing* - Several investigations have shown jobs involving all standing or all sitting postures were associated with increased risk for low-back pain as compared to jobs involving frequent changes in posture.

7. *Vibrations* - As in other CTDs, vibration is a significant risk factor.

8. *Accidents* - Traumatic events outside the context of manual lifting, such as slipping, tripping, stumbling, or other incidents which place unexpected loads on the back, can contribute to chronic low-back pain.

Personal Factors:

1. *Age* - Low back pain usually begins early in life, occurs greatest in persons aged 35 to 55 years, and has recurrences and increasing disability with older age.

2. *Gender* - Incident of low-back pain is equal among males and females. Women performing physically heavy jobs have significantly higher incidence of low-back pain injuries and a larger percentage of expensive workers' compensation claims; however, most workers' compensation claims (75 to 80%) are filed by men.

3. *Anthropometry* - No strong correlation exists between stature, body weight, or body build with low-back pain, although some studies show taller people to have a greater incidence of low-back pain.

4. *Physical fitness and training* - While it widely believed that physical training is associated with a reduction of low-back pain incidence, researchers disagree on what training provides the optimum benefits.

5. *Lumbar mobility* - There is no evidence that reduced spinal mobility is a risk factor for low-back pain.

6. *Strength* - Controversy exists for the relevance of lower back strength on the incidence of low-back pain. Some investigations show poor performance in physical tests such as sit-ups as related to higher risk of low-back pain.

7. *Medical history* - Individuals with a history of low-back pain are more likely to have prolonged disability associated with this illness and recommend against such individuals from engaging in jobs which require heavy physical work.

8. *Structural abnormalities* - Congenital or acquired skeletal defects have been found to give rise to low-back pain but other individuals with defects such as scoliosis, leg-length discrepancy, spina bifida occulta, and osteoarthritis do not show higher prevalence for low-back pain.

Relating low-back pain to a particular accident or incident can often be difficult. Many attribute the onset of back pain to a particular activity, but only a few can relate the onset to a particular incident. Other studies show that in one-third of lower back injuries, causation from one particular incident cannot be established. Furthermore, some studies indicate that about a quarter of patients do not have pain until several days after the incident.

In a specific study performed among New York City fire fighters,[3] the role of fire fighting activities was examined as specific risk factors for the incidence of low-back pain. Findings from this study included:

- Statistically significant *high-risk work activities* included operating a charged hose inside a building, climbing ladders, breaking windows, cutting structure, looking for hidden fires, and lifting heavy objects (> 40 lbs).

- Statistically significant *low-risk work activities* included connecting hydrants to pumpers, pulling booster hose, and participating in drills or physical training.

- When the review of incidents was adjusted to exposure to smoke as an indication of alarm severity, the odds for low-back pain associated with high-risk work activities were no longer significant.

- When the risk of low-back pain was measured for fire fighter proximity to the incident, the risk gradually increased as the fire fighter moved away from the fire house and closer to the site of the fire.

---

[3]Iman A. Nuwayhid, Walter Stewart, and Jeffrey V. Johnson, "Work Activities and the Onset of First-Time Low Back Pain Among New York City Fire Fighters," American Journal of Epidemiology, vol. 137(5), 1993, pp. 539-548.

Heat Stress

The body temperature must remain with a limited range or suffer life-threatening consequences. The center for temperature control is located within the hypothalamus region of the brain. If a rise in body temperature is detected, blood warmed by the inner core temperatures of the body is circulated to dilated blood vessels in the skin, where the excess heat can be given off to the cooler environmental temperatures. If this action is not sufficient, the sweating mechanism will be invoked. It is extremely efficient at liberating body hear- and cooling the body. Perspiration formed on the skin will evaporate generating a tremendous cooling effect.

Owing to rigorous physical and environmental demands, heat stress is   a natural manifestation for fire fighting environments.   If the cooling systems of the body are overwhelmed, it does not take long to experience heat disorders which can be the source of great discomfort or even death. The fire fighter or EMT is at great risk because all available means for dissipating heat are either hampered or rendered completely useless.

Types of Disorders. Heat stress disorders take a variety of forms ranging from mild to life-threatening. Mild forms of heat stress disorders are fainting, heat rash, and heat fatigue. Severe forms of heat stress are heat cramps, heat exhaustion, and heatstroke. A synopsis of these disorders is provided below.

SIGNS AND SYMPTOMS OF HEAT STRESS

- **Heat rash** may result from continuous exposure to heat or humid air.

- **Heat cramps** are caused by heavy sweating with inadequate electrolyte replacement. Signs and symptoms include:
  O muscle spasms   O pain in the hands, feet, and abdomen

- **Heat exhaustion** occurs from increased stress on various body organs including inadequate blood circulation due to cardiovascular insufficiency or dehydration. Signs and symptoms include:
  O pale, cool, moist skin   O heavy sweating
  O dizziness   O nausea
  O fainting

- **Heat stroke** is the most serious form of heat stress. Temperature regulation fails and the body temperature rises to critical levels. Immediate action must be taken to cool the body before serious injury and death occur. Competent medical help must be obtained. Signs and symptoms are:
  O red, hot, usually dry skin   O lack of or reduced perspiration
  O nausea   O dizziness and confusion
  O strong, rapid pulse   O coma

Risk Factors. Obviously the likelihood for heat stress is the consequence of workplace factors including:

1. *Environmental conditions* - risk increases as temperature and humidity increase. Several indices have been established on the basis for regarding safe working limits in different environmental conditions. The wet bulb globe temperature (WBGT) is one such index.

2. *Level of work activity* - higher work activities produce earlier onset of heat stress-related symptoms given the environmental conditions and work/rest cycles; Direct relationships exist between the level of work and environment conditions which can be tolerated by individuals.

3. *Type of PPE worn* - increasing amounts of clothing and the nature of clothing significantly affect the potential for heat stress. Fire fighters turnout clothing is often regarded as encumbering. The same characteristics which are used to protect fire fighters from outside environmental hazards also act to contain heat and prevent normal sweating to negate normal body mechanisms for losing heat. As clothing becomes more encumbering, such as in totally encapsulating chemical protective suit, these effects can be more pronounced.

Individuals vary in their susceptibility to heat stress. A number of factors can predispose fire fighters and emergency medical technicians for having heat stress, including:

1. **Physical condition** - Physical fitness is a major factor influencing a person's ability to perform work under heat stress. The more fit someone is, the more work they can safely perform. At a given level of work, a fit person, relative to an unfit person, will have:
    *   Less physiological strain;
    *   A lower body temperature, which indicates less retained body heat (a rise in internal temperature precipitates heat injury);
    *   A more efficient sweating mechanism;
    *   Slightly lower oxygen consumption; and.
    *   Slightly lower carbon dioxide production.

2. **Level of acclimization** - The degree to which a worker's body has physiologically adjusted or acclimatized to working under hot conditions affects his or her ability to do work. Acclimatized individuals generally have lower heart rates and body temperatures than unacclimatized individuals, and swat sooner and more profusely. This enables them to maintain lower skin and body temperatures at a given level of environmental heat and work loads than unacclimatized workers. Sweat composition also becomes more dilute with acclimatization, which reduces salt loss. When enclosed in an impermeable suit, fit acclimatized individuals sweat more profusely than unfit or unacclimatized individuals and may therefore actually face a greater danger of heat exhaustion due to rapid dehydration. Acclimatization can occur after just a few days of exposure to a hot environment.

3.   **Age** - Generally, maximum work capacity declines with increasing age, but this is not always the case. Active, well-conditioned seniors often have performance capabilities equal to or greater than young sedentary individuals. However, there is some evidence, indicated by lower sweat rates and higher body core temperatures, that older individuals are less effective in compensating for a given level of environmental heat and work loads. At moderate thermal loads, however, the physiological responses of "young" and "old" are similar and performance is not affected. Age should not be the sole criterion for judging whether or not an individual should be subjected to moderate heat stress. Fitness level is a more important factor.

4.   **Gender** - The literature indicates that females tolerate heat stress at least as well as their male counterparts. Generally, a female's work capacity averages 10 to 30 percent less than that of a male. The primary reasons for this are the greater oxygen-carrying capacity and the stronger heart in the male. However, a similar situation exists as with aging: not all males have greater work capacities than all females.

5.   **Weight** - The ability of a body to dissipate heat depends on the ratio of its surface area to its mass (surface area/weight). Heat loss (dissipation) is a function of surface area and heat production is dependent on mass. Therefore, heat balance is described by the ration of the two. Since overweight individuals (those with a low ratio) produce more heat per unit of surface area than thin individuals (those with a high ratio), overweight individuals should be given special consideration in heat stress situations. However, when wearing impermeable clothing, the weight of an individual is not a critical factor in determining the ability to dissipate excess heat.

Methods of Diagnosis. The principle methods of monitoring and diagnosing heat stress include:

- Individual complaints of cramps, dizziness, fainting, nausea;

- Skin condition (pale, cool, moist skin indicates heat exhaustion while red, hot, dry skin is evidence of heat stroke);

- Oral temperature (temperatures above 99.6°F modification of work practices while temperatures above 101°F are considered dangerous) ;

- Heart rate (continued, elevated high pulse rates can be related to the onset of heat stress); and

- Weight loss (NIOSH guidelines recommend that body water loss not exceed 1.5 % percent total body weight in a work day).

While most of these conditions may seem obvious, the attention of a physician or emergency medical technician is necessary to provide an accurate diagnosis.

Cold Stress

The major concern about whole body exposure to the cold is the development of serious hypothermia and subsequent death from exposure. The body defends core temperature by intense shivering to increase metabolic heat. Exhaustion of this resource for generating heat is implied when body temperature falls below 95°F. Frostbite of the face or extremities may result from exposure to extreme cold, often in combination with high air velocities, or from prolonged exposure to less severe cold but with high humidity. Exposures may result in cold injury to exposed flesh at equivalent temperatures of -32°F. For individuals working outdoors in the cold, body heat losses associated with high winds can be very significant.

The body's ability to compensate for heat loss is less than its ability to tolerate increased heat. As ambient temperature falls, cold discomfort increases rapidly, even when more clothing is worn. The insulation of clothing slows the fall of skin temperature below 91°F during light work in cold environments, and it reduces convective heat loss in high air velocities. Shivering is initiated at skin temperature of about 86°F, and this effect further discomforts the individual.

Local cold discomfort, most often in the hands, feet, and face, is usually the major cause of complaints in the cold discomfort zone. The hands begin to exhibit some loss of flexibility and manipulation skills at ambient dry bulb temperatures of 60°F over a few hours of exposure. A 20 % decrement in performance is not unusual in manual tasks at ambient temperatures of 7°F, dry bulb. Extended exposure to cold conditions leads to frostbite which in turn can lead to permanent injuries if left untreated.

Susceptibility to cold is primarily a function of the environmental conditions (temperature, humidity, and wind velocity), length of exposure, level of work activity, and amount of clothing worn.

Noise Stress

Many of the fire fighters' noise exposures are obvious, including the sirens and air horns that warn the public of the approaching emergency vehicles, as well as the roar of the diesel engines that power the trucks. As mechanized equipment replaces older hand tools, the equipment used on the fire scene is becoming louder. Even the living quarters of fire fighters can be noisy, being located on or near major highways for easy access to roads. The stations are often built with shiny, hard-surfaced materials that reflect the noise around the living quarters. Many kitchens have metal stoves and heavy metal cooking utensils that contribute to the overall noise emissions. Departments that have the additional responsibility of airport fire and rescue also hear the aircraft landings and take-offs.

Many of the tasks fire fighters perform depend on the auditory ability of the person. It is nearly impossible to see in a smoke-filled room. This loss of visual acuity is typified in the training adage of "go for the glow," which directs fire fighters to the area of the working fire. They are also taught to listen for moans and cries when conducting a rescue search. In addition, they must listen to and respond to radio communications as well as for the warning sound of an air horn that signals fire fighters to leave the interior of a building because of imminent danger. A loss of auditory acuity can literally be a life-and-death situation for people engaged in this type of work. These critical hearing requirements have led to a body of research that has quantified the hearing abilities of fire fighters and the amount of noise to which they are exposed in their occupation.

Excessive noise can affect fire fighters or EMTs in any of the following ways:

- It may contribute to hearing loss;
- It may interfere with communication;
- It may annoy or distract individuals; and
- It may alter performance on some tasks.

Each of these ramifications can have significant impacts on fire fighting and emergency medical operations.

Criteria have been established by several sources for defining levels of acceptable noise in the workplace:

1. *Hearing loss* - Although it occurs gradually, hearing loss represents irreversible damage to the inner ear. The degree to which hearing is affected depends on the intensity, frequency spectrum, and duration of noise exposure, plus individual susceptibility. Noise-induce hearing loss usually produces loss of the high frequency components first, resulting in reduced quality, clarity, and fidelity of sounds. Research has indicated that extended exposure, about 8 hours, to noise levels in excess of 85 dBA (decibels on the A scale of a sound-level meter) may cause hearing loss.

2. *Annoyance and distraction* - Although noise levels below 85 dBA probably do not contribute to hearing loss problems, they may contribute to performance decrements due to distraction or annoyance. Noise equipment can reduce the effectiveness of communications and make it difficult for people to concentrate in some types of tasks.

3. *Interference with communication* - Speech interference by noise is fairly common around noisy machinery or vehicles. The criticalness of communications should determine the steps to be taken to improve the noise levels in the environment.

4. *Impact on work performance* - Noise may affect performance in a number of ways. Noise features which are most likely to degrade performance include variability in level or content, intermittency, high-level repeated noises, frequencies above 2000 Hz, or any combination of these features.

Several studies of noise exposures have been conducted for the fire and emergency medical services. Most of the these studies have provided data characterizing noise exposures during emergency medical operations, some exceeded 120 dBa for short times. Noise measurements have generally been divided into four categories:

1. Code 3 operations;

2. fire scene environment;
3. return travel to station; and
4. station environment.

Fire station log-book records were used to estimate times at the sation and the number of emergency runs. The calculated noise levels suggest that fire fighters were exposed to noise levels that exceeded the OSHA Permissible Exposure Limit (PEL) of 90 dB(A). These excessive exposures were found for fire fighters occupying nearly all of the riding positions on the vehicles.[4]

Methods of Diagnosis. Audiometric test results are combined according to different criteria to ascertain the hearing handicap for speech perception or for the determination of noise effects on hearing. A criterion proposed by NIOSH in its criteria document for occupational noise exposure is intended to determine the amount of handicap in speech perception and communication abilities. It averages the hearing level, or amount of hearing deficit, in decibels (dB HL) at the pure-tone frequencies of 1,000, 2,000, and 3,000 Hz for each ear. The criterion incorporates a 25-dB "low-fence" value. This means that the dB HL average value must exceed 25 dB before a hearing handicap is said to exist. The percentage of handicap is calculated by multiplying each decibel in excess of 25 dB HL by 1.5 %. For example, an average dB HL of 40 for this metric would represent a 22.5% hearing handicap.

A second formula has been proposed by the American Academy. of Otolaryngology-Head and Neck Surgery. The criterion combines the pure-tone frequencies of 3,000,4,000, and 6,000 Hz and states that an average change of 20 dB at these frequencies will serve as an otologic referral criterion. The referral is for otologic evaluation to determine the nature and cause of the hearing impairment. This combination is most sensitive to the sensorineural effects on the ear from noise because of the propensity of these frequencies to deteriorate sooner when a person is exposed to loud noises.

Finally, a criterion proposed by Eagles et al. for single-frequency hearing impairment determination also uses a low fence of 25 dB HL. With this criterion, any person who has a hearing level of 26 dB HL or greater at any single frequency is classified as having some degree of hearing loss. The degree of loss could range from mild (26-40 dB HL) to profound (> 90 dB HL). This criterion differs from the other two criteria in that it looks at single test frequencies rather than average hearing levels across several frequencies.

Visual Stress

The use of computers is becoming more common. As a result, many people spend all or part of their workdays using a video display terminal (VDT). Eyestrain is the single largest category of complaints among VDT users. Eyestrain is often worst for personnel doing intensive work, looking at an interactive terminal all day or continuously looking back and forth between hard copy and the screen. Also, excessive overhead illumination that causes the glare on the

---

[4]Randy L. Tubbs, Noise and Hearing Loss in Fire Fighting, Occupational Medicine: State of the Art Reviews, Vol. 10(4), October-December, 1995, pp. 843-856.

VDT screen can result in eye irritation, eye fatigue, headaches, and blurred vision. Right now there is no evidence linking VDT use to any permanent visual damage.

Other visual stress can occur simply from poor lighting conditions or excessive glare. Fire fighters and EMTs often have little control over the environments where emergencies occur. The provision of sufficient illumination can mean the difference in fire fighters or EMTs successfully and safely performing emergency functions.

CHAPTER 3 - DEVELOPING AN ERGONOMICS PROGRAM

The goal of an ergonomics program is to make the workplace as adaptable as possible to the workers, and in the case of fire fighters and EMTs, allow them to safely function within the work environment. In order to achieve this goal, a successful ergonomics program for the fire/EMS department must consist of seven specific elements described in the box below.

ESSENTIAL ELEMENTS OF AN ERGONOMICS PROGRAM

1. **Assessment of Ergonomics Hazards.** Departments should know how to identify ergonomic hazards to which personnel are exposed. Observing specific tasks by using checklists and analyzing injury data are two common methods for assessing hazards. (See Chapter 4).

2. **Prevention and Control of Ergonomic Hazards.** Once hazards are identified, appropriate solutions must be devised to prevent and control personnel exposure through appropriate tactics, equipment/tool selection, and administrative controls, where possible. (See Chapter 5).

3. **Training.** Department personnel must be trained to recognize ergonomic hazards and follow prevention/control techniques to avoid injuries. (See Chapter 6).

4. **A Medical Management System.** Medical supervision is required to develop both preventative measures (e.g., fitness training) and reactive measures (e.g., rehabilitation). (See Chapter 7).

5. **Procedures for Reporting Injuries.** The department should have an injury reporting system that allows its to spot trends in injuries which indicate ergonomic hazards. (See Chapter 8).

6. **A Plan for Implementation the Program.** Specific individual responsibilities must be established for starting and maintaining the program. (See Chapter 9).

7. **Methods for Evaluating Program Effectiveness.** Periodic review is needed to determine how well program reduces injuries. (See Chapter 10).

Specific guidelines for each of the program elements are provided in separate chapters as indicated above.

General Steps for Program Development

The following steps are recommended for developing an ergonomic program for the fire and emergency medical services:

1. Gain an understanding of ergonomics and how it applies to your fire and/or EMS department.

2. Secure the commitment of senior department officials or local government to go ahead and develop a plan, Consider using the cost/benefit analysis approach described in **Chapter 10** and **Appendix G** to provide justification for the plan. Consider developing your plan so that it becomes part of your existing occupational safety and health program.

3. Assess the ergonomic hazards in your department by job and task. Identify those jobs and tasks which involve the greatest risk of injury from ergonomic hazards.

4. Determine and evaluate potential solutions to eliminate or reduce ergonomic hazards.

5. Integrate ergonomics-based training into existing department safety training.

6. Establish a rigorous medical management system which addresses both preventative measures and treatment protocols for all personnel. An important part of this system is the establishment of a fitness training program with periodic fitness evaluations.

7. Develop improved procedures for reporting injuries which allow your department to discern between injuries due to ergonomic hazards and those caused by other factors. Such a injury reporting system will have widespread benefits for management and your occupational safety and health departments in general.

8. As with any program, ensure that its implementation is complete and periodically evaluate its effectiveness.

Appendix C offers a sample ergonomic program plan as a starting point. Your department's specific ergonomic plan should be include details relevant to your department's activities and be consistent with available resources for implementing and operating the program. Any program which is developed should meet two important criteria:

1. The program should be effective in reducing ergonomic hazards as evidenced by lower injury rates, increased response efficiency (better service), and improved morale; and

2. The program should designed to operate within the available staffing and budget for the department (although adjustments to both staffing and budgets should be considered if necessary to produce the desired benefits).

Relevant Standards and Key Resources for Developing an Ergonomics Program

NFPA *1500, Standard on Fire Department Occupational Safety and Health Program,* establishes minimum requirements for fire and emergency service department occupational safety and health programs. As such, ergonomic considerations are just one facet of health and safety concerns for this overall program. Therefore, *the department ergonomics program should be viewed as a specific part of the department occupational safety and health program prescribed in NFPA 1.500.* Key elements of NFPA 1500 which pertain to development of an ergonomics programs (and included in this guide) are shown in the box below.

KEY ELEMENTS OF NFPA 1500 WHICH PERTAIN TO THE DEVELOPMENT OF AN ERGONOMICS PROGRAM

Section 2-2 Risk Management Plan: prescribes general approach for assessment/control of hazards with specific recommendations in Appendix.

Section 2-3 Policy: prescribes adoptions of safety and health goals for prevention and elimination of accidents; also sets frequency for evaluating effectiveness of occupational safety and health program.

Section 2-4 Roles and Responsibilities: prescribes establishing set roles and responsibilities for specific individuals in department for carrying out program.

Section 2-6 Occupational Safety and Health Committee: prescribes committee responsible to advise department management on safety and health issues, review safety and health concerns, and to hold/document regular meetings.

Section 2-7 Records: prescribes departments to have systems for reporting and recording injuries.

Chapter 3 Training: prescribes regular training of personnel on safety and health issues.

Chapter 4 Vehicles and Equipment: prescribes safety and health considerations for specification/selection of apparatus and tools.

Chapter 6 Emergency Operations: prescribes awareness of environmental conditions and on-scene rehabilitation.

Chapter 8 Medical and Physical: prescribes regular medical examinations, physical fitness programs, and post-injury rehabilitation.

The provisions within NFPA 1500 should be examined during the development of an ergonomic program to determine how it should be structured to fit into the department's occupational safety and health program.

In addition to NFPA 1500, a number of key resources are available to assist you with the development of an ergonomics program. These include other NFPA standards and publications prepared by the U.S. Fire Administration which offer guidance and findings in specific areas. **Appendix A** includes an annotated bibliography which lists a number of articles and books prepared on the subject of ergonomics, some of them directly related to the fire and emergency medical services. **Appendix B** lists general sources of information on ergonomics and key government publications.

OSHA Draft Proposed Ergonomic Protection Standard. The Occupational Safety and Health Administration (OSHA) released a rough draft of a proposed ergonomic standard in March 1995. The development of this proposed standard was undertaken to target industries which have typically experienced high levels of cumulative trauma disorders. The draft was introduced to get information from companies and organizations concerning the feasibility and practicality of such a standard since OSHA wanted to ensure that the standard would be beneficial to industry. Specific provisions of this proposed draft standard are provided below.[2]

SUMMARY OF PROVISIONS WITHIN PROPOSED OSHA DRAFT ERGONOMICS STANDARD

- organized in to sections on scope/application, methods of control, potential control measures, requirements for training, medical management, and record-keeping.

- intended to apply to any workplace where certain risk factors exist (e.g., repetitive motion, awkward work posture, use of vibrating tools, and unassisted manual lifting),

- requires employer to identify 'problem' jobs involving risk factors; recommends use of checklists for ranking 'problem' jobs.

- requires employer to control problem jobs by implementing engineering or administrative controts.

---

[2]At the time the final draft of this guide was prepared (March 1996), the adoption of an OSHA-based ergonomic standard was considered unlikely for next several years.

Other Key References. References used by OSHA in developing its proposed ergonomics standard are available and may provide further help in developing an effective program:

- OSHA Draft Proposed *Ergonomic Protection Standard: Summaries, Explanations, Regulatory Text, Appendixes A and B* published March 20, 1995 in the Occupational Safety and Health Report, Bureau of National Affairs, Special Supplement.

- American National Standard Institute (ANSI) Z-365 *Working Draft Control of Work-Related Cumulative Trauma Disorders, Part 1: Upper Extremities* published July 14, 1994. This document was developed by a committee of ergonomic experts which was chartered under the direction of ANSI to develop a voluntary ergonomics standard for industry that OSHA could potentially use in its standard-setting process.

- OSHA's *Ergonomics Program Management Guidelines for Meatpacking Plants.* This document was considered a possible blueprint for an OSHA standard because much of the information for ergonomics in the meatpacking industry also pertains to general industry, including fire fighting and emergency medical operations. The guidelines outline the four major program elements of an ergonomics program that OSHA recommended in its proposed draft standard: worksite analysis, hazard prevention/control, medical management, and training.

- OSHA's *Advanced Notice of Proposed Rulemaking (ANPR) on Ergonomic Safety and Health Management,* published in the Federal Register, Monday, August 3, 1992. This is another document that gives insight into OSHA's thinking. The ANPR asks many more questions than it answers, but is the first formal step that OSHA must taken when setting a standard and it is self-explanatory. The idea is to put industry on notice that a standard might be coming, and it asks for public comment on the topic.

- California's proposed general industry standard for the prevention of cumulative trauma disorders (CTDs). In November 1993, the state's Occupational Safety and Health Board released a proposed ergonomics standard. The proposed California standard covers all employers, despite size, and asks them to review safety, workers' compensation, medical and OSHA "200 log" records for evidence of CTD or other risks. Employers must develop a procedure for encouraging workers to report CTD symptoms. Worksite evaluation must be done whenever CTD problems or risks surface through employee injury reports. These evaluations can focus on one employee's job, or a group of activities, depending on the nature of the problem. Guidelines are included for assessing risks. workstation changes, administrative controls such as rest periods, or Personal Protective Equipment (PPE) must be used when a job causes, aggravates, or is likely to cause any CTD symptoms.

Medical evaluations must be provided whenever an employee reports a CTD symptom. A written medical evaluation must be supplied to the employee, if requested.

Within one year of the California standard being issued, employers must provide ergonomic training to all employees. Workers must be made of the symptoms and consequences of CTDs, the type of job that can cause CTD risks, and safe work practices to reduce any associated risks.

Before attempting to enact any of these elements, OSHA strongly recommended that management be committed to setting up an ergonomics program and improving the safety and health of employees. This is an essential step in the implementation of the program and is described in **Chapter 9.**

Learning to Recognize Ergonomic Problems and Problem Areas

Each job or task has its own set of actions, and each job or task has its own level of stress. In fire fighting and emergency medical operations, this stress can take on a number of forms and have different effects on both the health and performance of the fire fighter or EMT as shown below.

TYPES OF JOB/TASK STRESS

Stress Type	Health Effects	Performance Effects
Physical	Muscle, joint pains Cumulative trauma Sensory loss Reduced strength	Interference with manipulative tasks
Heat	Muscle cramps Heat illness Circulatory collapse	Distraction Interference with manipulative tasks
Cold	Shivering Hypothermia Frostbite Loss of flexibility in extremities	Distraction Interference with manipulative tasks
Noise	Fatigue Hearing loss	Interference with communication and signal detection
Visual	Fatigue Headache Reduced visual acuity Eye injury	Distraction Reduced detection of hazards

Physical stress includes those factors which relate the level of exertion and orientation (or position) required to carry out a specific motion or activity. Physical stress is therefore caused by working with excessive weight and working in awkward positions. The frequency of these tasks and associated factors such as vibration affect the likelihood that these stresses will cause harm to the fire fighter or EMT.

Environmental stresses such as heat, cold, noise, and visual difficulties can affect fire fighters or EMTs in of themselves or combined to produce greater stress on the individual. The severity of these stresses and their frequency must be determined when assessing ergonomic hazards.

Recognizing ergonomic hazards in the workplace is an essential step in correcting them. Other issues that must be considered when evaluating physical stress can be outlined as follows:

- Does the fire fighter or EMT sit or stand to do the particular task? Does the task require a variety of positions? Are any of these positions awkward or involve unnatural/difficult body positions for any length of time?

- Is the fire fighter or EMT stationary when doing the job, or must he or she move around?

- Does the job require a great deal of strength or power?

- Do most fire fighters or EMTs have a comfortable reach, or must they stretch and bend excessively?

- Is the work at a comfortable height, or must the worker perform the job in awkward positions (bent over from a too-low work height)?

- Are job tasks extremely repetitive?

- Do fire fighters or EMTs have any control over the pace of the job?

- Are there elements in the job environment that can cause discomfort, injury, or illness (extreme temperatures, noise, improper lighting, etc.)?

Major causes of current problems are more specialized tasks, uncontrolled conditions, increased repetition, and a lack of ergonomically designed technologies. Consequently, fire fighter and EMT hands, wrists, arms, shoulders, backs, and legs may be subjected to excessive repetitive twisting and forceful or flexing motions during a specific response or workday. Some jobs still expose individuals to excessive vibration and noise, eye strain, repetitive motion, and heavy lifting. Commonly, machines, tools, and the work environment are often poorly designed, placing undue stress on tendons, muscles, and nerves. In addition, temperature extremes common on the fire ground may aggravate or increase ergonomic stress.

Although a great variety of symptom survey and risk analysis forms have been developed by universities, institutes, and ergonomic consultants, the basic process for assessing ergonomic hazards follows the sequence outlined below.

1. Identify and prioritize jobs or tasks where:
   work-related injuries have been known to occur
   a. personnel have been restricted from some parts of a job after an injury or illness

c. personnel have left the job because of inability to perform or dislike of either environment or physical requirements

d. older or smaller fire fighters or EMTs have limited success doing the job or staying on it

e. injury-free personnel have above-average strength, fitness, and endurance capabilities

f. people have difficulty sustaining quality performance

g. it takes more than three months to become proficient on the job

2. List problems and classify by type/job

3. Determine which jobs are mentioned under item 1 above

4. Perform analysis of jobs selected in item 3

Identification of High-Risk Jobs

An essential element in developing an ergonomic program is assess the specific ergonomic hazards faced by the department. A detailed department screening survey must be conducted to identify high-risk jobs or tasks with the potential for causing CTDs and other ergonomic-related injuries. The purpose of the survey is to identify jobs/tasks that place personnel at risk of developing CTDs.

At a minimum, the following factors should be included in the ergonomic assessment of specific jobs or tasks, as applicable:

- Upper extremities that are used; total hand manipulations per task (frequency of basic operation in the task), and total hand manipulations per average response;
- Average force or effort (light, moderate, or heavy);
- Tools being used;
- Sources of vibration;
- Types and use of personal protective equipment (PPE);
- Levels and adjustability of environmental conditions; and
- Levels of risk for hand, arm, and shoulder postures.

The next step is evaluate the work (or response) environment. Because emergency scenes vary from response to response, studying every aspect of the environment may not be necessary, but looking for general conditions will indicate whether an ergonomic approach is needed to improve control and lessen injuries. For example, if during overhaul procedures, a new tool is introduced for pulling down ceilings, ergonomics planning should be an important part of the process. As with any hazard assessment, there are some indicators that must be reviewed:

- Trends in accidents and injuries, incidents of CTDs and other ergonomic stress-related disorders;
- Frequent personnel complaints;
- Personnel-generated changes in the workplace (tools or equipment modifications);
- Poor response times (because of awkward or hampered conditions);

- More than necessary manual material handling and repetitive motion tasks; and
- Work done by personnel with reduced capabilities.

These are only some of the more common indicators. Other indicators might be present based on specific department practices and procedures, e.g., types of hydraulic extrication equipment used.

The purpose of analyzing jobs and tasks is to recognize and identify ergonomic hazards so that they can be corrected. The analysis involves reviewing injury and illness records, insurance records, and medical records to determine rates of various ergonomic incidents and injuries. Trends can then be evaluated as they relate to ergonomic hazards by:

- departments (divisions or battalions),
- types of operations/responses,
- responder roles, and
- specific tasks.

Worksite analysis identifies problem tasks and the risk factors associated with them. This essential preliminary step helps determine which tasks and the environment conditions which are the source of the greatest problems. The most effective worksite analyses will include all jobs, operations, and work activities where there are ergonomic risk factors, regardless of whether the department's medical records indicate ergonomic illnesses. A review of commonly reported symptoms is also a clue as described below.

EXAMPLE FOR HOW SYMPTOMS SHOW ERGONOMIC HAZARDS

Video display tube (VDT) operators (e.g., fire and EMS dispatchers) have experienced eyestrain; headaches; excessive fatigue; neck, back and muscle pain, and stress. Research has shown that these injuries are often the result from problems associated with the communications center environment.

Additional Methods for Identifying Job or Task Hazards

There are several ways to identify job or task hazards:

- Evaluate the ways in which human error might contribute to the hazards of a particular job;

- Record the types of potential accidents and the physical agents or environmental conditions involved:

- Ensure that procedures are clearly written so that the job or task is outlined in a simple form; and

- List all of the jobs or tasks in the specific area being investigated.

Once the jobs and tasks have been identified and the basic steps outlined, hazards can be more easily identified, Each step should be evaluated to identify all real hazards, both physical and physiological. If the task requires heavy lifting, twisting, pushing or pulling, it should be especially noted. Questions to be asked when identifying hazards include:

- Are there sharp edges?

- Is there danger of striking against, being struck by, or contacting a harmful object?

- Can an individual be caught in, on, by, or between moving and stationary objects? Body areas, parts of clothing, or equipment may be pinched, crushed, or caught between either a moving object and a stationary object, or two moving objects.

- Is there a potential for a slip, trip, or fall? Can personnel fall from the same level or a different level? People can slip, trip, or fall to the surface on which they are standing or walking. They also could fall to a level below the one on which they are standing or walking.

- Can individuals strain themselves by pushing, pulling, lifting, bending, or twisting while performing this job/task? Individuals can also hyperextend a joint or strain their backs by twisting and bending.

- Is the environment hazardous to safety or health? Are personnel can be exposed to harmful substances or conditions?

- What are other hazards, not classified above, having the potential to cause an accident? Note equipment that is difficult to operate and that could be used incorrectly can possibly cause injury. All equipment should be in proper working condition. Determine what stress level the individual is experiencing via some objective or subjective technique.

Before a task analysis program is started, it is important to establish good lines of communication with everyone involved, This will reduce the natural tendency of personnel to be suspicious of someone who suddenly arrives at the workplace with cameras, notebooks, and asking many questions. Introducing the program properly can also give you the benefit of the fire fighter's or EMT's insights. Before the task analysis program is started, these questions should be answered for everyone involved:

- What are the goals and intent of the ergonomics analysis?
- Who will be involved?
- Exactly what is going to happen?

A decision should be made as to the best way to inform the personnel within the department about the job/task hazard analysis. Small group meetings, bulletin board notices, and memos are all good ways to get the information out to people. It is important to tell personnel that task analysis can pinpoint the following problems or needs:

- work tasks that involve (adverse) movements,
- tasks involving (excessive) manual lifting,
- wasted motion,
- work tasks that do not permit comfortable posture,
- tasks that present the potential for psychological stress,
- fatigue, and
- tasks that have or possess controllable hazards.

A good filing system may be needed to keep track of the results of job evaluations, ideas for improvement, planned changes, and overall progress. The system can range from a simple notebook or file drawer to a computer spreadsheet or a database system.

Determination of Priority Jobs/Task for Analysis

When assessing ergonomic hazards in the department, the following points should be considered in setting priorities:

- *Job injury and occupational illness* severity. Those jobs that have involved serious accidents or injuries should take priority. There may be a basic problem in the work environment or in how the job itself is performed.

- *Job accident frequency.* The higher the frequency rates of accidents, the greater the reason for implementing a job-hazard analysis.

- *Potential for illness or injury,* even if no such incident has occurred.

- A *new job/task* for which there is no accident history or information about its potential for accidents or injuries. Many accidents occur in a job or task to which a fire fighter or EMT is unaccustomed; an analysis of the job can uncover hazards, and reduce or eliminate current hazards.

Table 4-1 provides a number of examples of fire fighting and EMS tasks with ergonomic hazards.

Job-Hazard Analysis

The surveys in evaluation of job and task environments together with the identification of job/task hazards will result in a list of jobs and tasks which are possible high-risk, plus a summary of the specific hazards associated with those jobs/tasks. Jobs identified as high risk should be further evaluated. This second analysis is sometimes called a job-hazard analysis and includes a much more detailed analysis of ergonomic hazards. The job-hazard analysis should be performed by a qualified person such as an ergonomist or someone in the department who is experienced with risk assessment from an ergonomics perspective.

Table 4-1. Common Fire Fighting/EMS Activities Involving Ergonomic Hazards

HIGH-RISE FIRES
• Moving equipment and high-rise packs to the fire floor while wearing full turnout gear • Moving additional equipment (tools, extra air tanks, etc.) to staging area immediately below the fire floor
VENTILATION AND OVERHAUL PROCEDURES
• Breaking through a roof while on a ladder or a pitched roof • Using a pike pole to pull down a ceiling
HOSE LAYING OPERATIONS
• Dragging a charged hose through a fire site, both inside (hallways/stairs) and outside (obstacles/icy conditions) • Directing a hose for an extended period of time • Laying a hose to a fire site from a distant hydrant
LADDER WORK
• Rescuing a victim from a roof or window using a ladder • Raising ladders while under-manned • Using an axe while on a ladder
FORCIBLE ENTRY
• Entry through steel security doors using hand tools • Using hand tools and power equipment to open a wall
EXTRICATIONS
• Using hand and power tools in confined areas to extricate victims • Using heavy hydraulic equipment in auto extrications • Moving victims from a damaged car or collapsed building • Moving and salvaging furniture
EXTENDED PROCEDURES
• Fighting fires for extended time period and conducting lengthy extrication procedures (automobile crashes, industrial fires, train derailment)*
EMERGENCY MEDICAL OPERATIONS
• Carrying first response kits from ambulance or apparatus to accident scene • Moving victims from multistory buildings or homes using stairways • Moving patients onto and off stretchers/gurneys and loading stretchers/gurneys into ambulances

When a job-hazard analysis is undertaken, direct department officer support must be available. Personnel become more receptive to changes in their job procedures when they are given an opportunity to help develop the change, and that improves safety awareness. Further, department personnel are the best source of on-the-job information, and appreciate being consulted on matters that affect them. Once the hazards have been identified, the correct solutions can be developed to protect the personnel from physical harm.

Benefits of a Job Hazard Analysis

Benefits of a job-hazard analysis go beyond safety. The job-hazard analysis provides actual step-by-step safety procedures for doing each task. This can be developed into a useful training tool. A job-hazard analysis program takes time, both to document and to implement effectively, and is a continuous process. The job-hazard analysis benefits a department because it encourages teamwork between supervisors and line-personnel and helps personnel realize that safety is a concern.

Job-hazard analyses allow supervisors and operators to identify risks together. A job-hazard analysis will help them to see the relationship between their job and safety; it also puts them in a position of control. Because a job-hazard analysis is developed collectively within a department, a sense of ownership is created, and operators are provided with a consistent method to do their jobs.

Detailed Methods of Assessing Hazards

A trained supervisor works with the fire fighter or EMT to record each step of the job as it is performed; consults with the fire fighter/EMT to identify hazards involved in each step; and, finally, enlists the fire fighter/EMT's participation on how to eliminate the hazards observed. The support of senior management makes job-hazard analyses easier to implement. Usually, the employee is the best source for identifying hazards. Involvement with employees and supervisors, where safety is developed as a team is crucial.

Techniques for conducting the job-hazard analysis include:

- general observations,
- questionnaires and interviews,
- measuring the work environment,
- breaking down the job,
- video analysis,
- photographs, and
- drawings and sketches.

The ergonomic job analysis attempts to relate the injury to the level of effort and other factors, usually by body part, using some arbitrary ranking of level of effort. A well-known job analysis form is shown on the next page (Figure 4-1). This form has the observer rate activity by body part in terms of effort, duration, and repetition using three point scales. As a result of these ratings, the priority is set as moderate, high, or very high. Since this form is primarily developed for analysis of precise job functions most often found in industry, it may be difficult to apply to many fire fighting and EMS situations as specific tasks changes with the response environment.

Knowing how to develop a job-hazard analysis to comply with ergonomics is not necessary. A job-hazard analysis is not a mandatory standard and employers are not required to use the idea; it is merely a good management tool. OSHA advocates the use of job-hazard analysis and has developed and issued Pamphlet 307 1, "Job Hazard Analysis," as a guide.

General Observations. Observation is particularly useful when reviewing jobs with skill level requirements and jobs with short repetitive cycles. These jobs can also be reviewed by interviewing the workers or simulating the tasks involved. Look for work tasks or situations that produce repetitive motion. Observation of the worker, work tasks, and the work environment are common and useful methods of obtaining job analysis data. Observation is particularly useful for studying jobs with low-level skill requirements and jobs with short, repetitive cycles. An experienced person can spot issues and potential improvements on a simple walk-through.

Questionnaires and Interviews. Once the job has been observed, it's a good idea to obtain the fire fighter's or EMT's participation. They usually know the job better than anyone since they do it all the time. This is the most important input into the ergonomics program that management can receive. In order to obtain their opinions, personnel can be surveyed by questionnaire or personal interview. Questionnaires make it easier to tabulate results but may not reveal the whole picture. The person analyzing the data must be able to judge the best approach for a given work environment. If questionnaires are used, guide the person through the questions, making sure the answers are consistent with the intent of the questionnaire (also watch for biases). This approach will give maximum validity to the results.

Another helpful approach in identifying a problem job is to ask people where they have experienced problems. This can be done both on an individual basis or in group meetings. Both supervisors and product employees usually have a good understanding of which tasks needs improvement. More formal survey techniques also can be valuable. One approach is to administer an anonymous personnel discomfort survey. Often, such surveys can provide valuable information about which jobs or tasks are causing what types of problems. A good survey can permit qualifications of discomfort so that work areas can be compared and before-and-after studies made. There are a variety of ways to administer surveys, depending on the needs and circumstances of the department.

Ergonomic Job Analysis

Body Part	Effort Level	Continuous Effort Time	Efforts/ Minute	Priority	Effort Categories
Neck/Shoulders	___	___	___	___	1 = Light 2 = Moderate 3 = Heavy
Back	___	___	___	___	Continuous Effort Time Categories 1 = <6 secs 2 = 6 to 20 secs 3 = >20 secs
Arms/Elbows	___	___	___	___	Efforts/Minute Categories 1 = <1/min 2 = 1 to 5/min 3 = 5/min
Wrists/Hands/ Fingers	___	___	___	___	
Legs/Knees	___	___	___	___	
Ankles/Feet Toes					

Priority for Change

Moderate =	1 2 3 1 3 2 2 1 3 2 2 2 2 3 1 2 3 2 3 1 2	Job Title: _____ Specific Task: _____ Job Number: _____ Department: _____ Location: _____
High =	2 2 3 3 1 3 3 2 1 3 2 2	Contact Person(s): _____ Phone: _____
Very High =	3 2 3 3 3 1 3 3 2	Analyst: _____ Phone: _____ Dale of Analysis: _____

Figure 4-1. Example Ergonomic Job Analysis Form

Breaking Down the Job. Measurements are an important part of workplace ergonomic analysis. Measurements, such as length of reach, work heights, task frequency, energy expenditure, etc., all describe a task. To say that a particular object is too heavy to lift or too hard to push depends upon these task measurements. Many think of fire fighter or EMT strength as the main criterion for work capacity. However, in jobs requiring highly repetitive tasks, the energy expended to perform these repetitions may be the limiting factor, not the worker's strength. Energy expenditure can be estimated by measuring oxygen consumption rate, but it is not usually done in the work environment because it requires complex equipment and special skills. Overall, people can work for long periods without undue fatigue if they work at one-third of their maximum capacity. Physical fitness is measured in terms of an individual's ability to consume oxygen.

Before hazards can be defined, the job must be broken down into individual tasks. Each task must be documented. When defining the job selected, keep in mind that most jobs consist of several different tasks. Every task that is a hazard requires a separate job-hazard analysis. No basic step should be omitted. However, care should be taken not to make the job hazards too detailed. Ensure that only "safety steps" are outlined. One of the most common mistakes is to mix work elements with job hazards. A job-hazard analysis is *not* intended to document work process instructions, although some people believe that they should be included. Too much detail will make a job-hazard analysis ineffective and unenforceable.

Ideally to evaluate a job effectively, one should be experienced, trained as to the purpose of a job-hazard analysis, have an open mind, and be provided with examples of correct methods. Focusing on safety is essential to the job being evaluated. To learn the components of the job, you need to understand both the basic steps needed to perform the task and the order in which they are performed. Each step must be documented as it occurs. This procedure should be reviewed several times to ensure that all the vital steps are documented.

Video Analysis. Videotaping workers at their jobs provides an excellent mechanism for evaluating body position, stress point, and repetitiveness. Slow motion and stop action give additional advantages for evaluation. Since the actual tasks are performed in rapid motion, direct visual observation is difficult. Videotaping is an excellent method of documenting the evaluation and timing the duration of exposure. When selecting a video camera and recorder/playback unit for use in ergonomic analysis, try to find one that can be used in low light. It is best if the camera is lightweight, portable, and provides sound recording.

Photography. Photographs and slides are all good mechanisms for analyzing a job. A motor-driven 35 mm camera can make it possible to break a task down into its elements. You can avoid distracting flash or other lighting by using a high-speed film. Instant photos and instant slides can be excellent tools when time is a factor.

Drawings and Sketches. A rough sketch or plan view of the workplace can contain selective details that may not be obvious in a photograph or shop drawing. Always be sure to measure and include dimensions and distances. A simple drawing of the work process and work environment might be all you need to work with for later evaluation of the work process.

Environmental Measurements. Some ergonomic stresses can be directly measured.

This is true for heat, cold, noise, and visual stresses:

- *Temperature, Humidity, and Velocity.* Common measurements include dry bulb, wet bulb, relative humidity, and air velocity at the response area. Sometimes these measurements may be useful to certain types of response such as hazardous material releases. Measurements should be taken of ambient surrounding whenever heat or cold stress problems are anticipated.

- *Noise.* Two types of noise measurement devices can be used. The standard sound-level meter measures the sound pressure in decibels, A more sophisticated sound-level meter is able to measure sound pressure against specific frequency bands. Noise levels should be measured in the apparatus, station, and at response scenes where excessive noise exists.

- *Illumination.* Light meters or photometers can be used to measure illumination levels. Normally response activities do not allow for measurement of illumination; however, these concerns can exist at the station and in other fire department facilities.

Analyzing Data, Recommending Changes, and Developing Solutions

Assemble the information and evaluate the data to determine if an ergonomic problem exists, and if so, to what degree. After the information is evaluated, recommendations can be made to management. One way of making recommendations could involve classifying ergonomic problems as follows:

- Problems that need immediate attention, i.e., health and safety hazards.

- Problems that present potential health and safety problems. These situations should be acted upon, but action does not have to be taken immediately.

- Situations that don't present an immediate risk to health or safety, but resolving them could result in productivity increases and increased employee morale. These situations would benefit from redesign or from the addition of new work space or equipment. The benefits might not be obvious, but should be given at least some consideration.

When it is determined that changes need to be made in certain jobs, recommendations to management should be clear and include information about the problem and suggestions for solution.

The final part in developing a job-hazard analysis is to determine the appropriate solution for each job hazard. This process should include the following steps:

- Determine if the task can be performed in a different, possibly less hazardous way (a possible solution would be to develop an entirely new method to accomplish the task).

- Redefine the goals to be accomplished and review the various methods of achieving those that will reduce accidents.

- Change the method of doing the task by changing the physical condition that creates the hazard.

- Modify the work process or procedures.

Finding ways to improve jobs is the key to any successful ergonomics program. An increasing amount of information is becoming available about the kinds of improvements that are possible. For example, you can list specific equipment and tools that can be reduce injury. If new methods are needed to reduce or eliminate hazards, a list of specific alternatives must be documented in order for the employee to know how to perform the job safely.

According to your observations and findings, recommendations of realistic changes must be documented to reduce the hazard or to eliminate accidents altogether. You may require a change in work procedures, policies, or perhaps a change in the entire system. Remember that all recommendations must be attainable. Change will come slowly. Use of persuasion, tact, and appropriate management language and perspective is critical when recommending potentially sweeping changes.

It should be recognized that ergonomic improvements are not always absolute; a period of experimentation, trial, and error is often needed to find the correct modification for the particular job task. The idea of continuous improvement is important. A job improvement is planned, implemented, evaluated, and redefined in an ongoing process.

If an incident occurs following a job-hazard analysis, review the analysis to decide if it needs revision. If the job-hazard analysis is revised, all employees involved in the job must be re-trained. If an accident resulted from misuse of a job-hazard analysis, all factors must be discussed with the employee. The supervisor, the fire fighter or EMT, and the ergonomics or safety committee should try to come to an understanding of why the procedure did not work. Administrative controls may be required.

Further Use of Job Hazard Analyses

To be effective, the job-hazard analysis information must be integrated into the operation by training all employees. Keep the job-hazard analyses available for each employee to use as needed. Supervisors are key players when training personnel; their involvement will help employees learn more about the jobs. Supervisors must explain the purpose and the content of the job-hazard analysis when it is distributed. Reviewing the entire document with personnel is important, as well as asking for feedback and comments. Any training should be documented permanently in the employee's records.

The analysis can be used to:

- train personnel about the hazards of the jobs, and
- increase safety awareness.

Instruct all personnel in the basic job steps using the job-hazard analysis as a guide. They will be taught to recognize the hazards involved in each step and reduce the probability of accidents. They will also be informed of the necessary precautions to take when doing the job.

The job-hazard analysis should only outline safety-related facts of the job and should not be used to detail the job. Often, mistakes are made where there is a mixture of job-hazard analysis and job instructions. This is often confusing to personnel. The only intent of a job-hazard analysis is to promote safety on a particular job.

The job-hazard analysis should be reviewed frequently with all personnel to ensure that injuries are minimized. Always solicit responses from the fire fighter, EMT, supervisor, and safety personnel. A job-hazard analysis is a good tool to help reduce accidents. An overall department objective should be to develop job-hazard analyses for every job.

CHAPTER 5 - HAZARD PREVENTION AND CONTROLS

Designing hazard controls into an ergonomics program is the major element toward solving workplace injuries. The goal is to engineer hazards out of the workplace by properly designing workstations, work methods, or tools to reduce trauma to the body. For the fire and emergency medical services, the extent of workplace engineering is likely to be limited given the changing nature of the response scene. Nevertheless, there are a number of areas which can be modified from fire apparatus to tools to response tactics (referred throughout as job tasks).

As has been discussed in previous chapters, the redesign of job tasks and the workplace begins with an analysis of the work environment. It's usually not necessary to use highly sophisticated methods. Even a cursory review of existing records and a rudimentary inspection of the workplace can quickly determine whether controls need to be selected and implemented.

TYPES OF ERGONOMIC CONTROLS

There are three basic types of ergonomics controls:

1. **Engineering controls** - changing the work place or equipment used in the work place to reduce or eliminate ergonomic hazards.

2. **Administrative controls** - limiting the individual's exposure to ergonomic hazards by changing their activity.

3. **Work practice controls** - modifying the task so that ergonomic hazards are reduced or eliminated.

Each of these controls are further discussed in the sections below.

Engineering Controls

Engineering hazards out of the job is the preferred method of control. This is because the primary focus of ergonomic hazard abatement is to make the job fit the person, not force the person to fit the job. Engineering controls are generally classified as three types:

1. workplace design,
2. work method design, and
3. tool design.

Workplace design adapts the work environment to make the fire fighter or EMT more comfortable, e.g., adjust the step on an fire apparatus to allow fire fighters easier access to equipment. This means that present equipment in the fire/EMS environment may have to be adapted to comply with ergonomic principles. All new equipment must be reviewed in the design phase to ensure that the appropriate ergonomic guidelines are followed. Design of work methods is the second type of engineering control. For example, personnel can put handles on boxes; provide bins; or fabricated boxes with built-in handholds so that the employee can easily lift the load. The third type of engineering control is tool design. An example is to provide counterbalancing for a tool that weighs more than two pounds.

When possible the work environment must be designed to accommodate a full range of required movements among personnel. The area should fit the body size and variability of personnel who are doing the job, It should also permit the fire fighter or EMT to adopt several different, but equally safe, postures that still permit performance of the job. Sufficient space should be provided for the knees and feet. Apparatus and other response equipment controls should be reachable and equally accessible by both right and left-handed operators.

To minimize back injuries, work locations should allow right- and left-handed personnel to lift loads with both hands while the chest faces the load. The workplace also should avoid the need for carrying objects overhead or too far away from the body. For example, workstations should be designed so that heavy loads are raised at least 18 inches off the floor before being lifted.

If the hazard cannot be controlled through engineering, the next options are administrative or work practice controls. These two methods are not as reliable as engineering controls because administrative controls need constant monitoring to ensure they are being followed.

Administrative Controls

Administrative controls are an important element of an ergonomics program since they are enacted or approved by management. They may act to reduce the duration, frequency, and severity of exposure to hazards. Frequent breaks or job rotation can be provided so that repetitive hand and body movement are reduced. The following examples show some administrative controls that can be used:

● Use rest breaks to relieve fatigued muscle groups or to reduce potential for heat stress. The length of time between breaks and the duration of the breaks depends on the task, work rate, and environmental factors.

● Increase the number of personnel assigned to the task. *Adequate staffing is a critically important consideration for lowering personnel injury rates as the work is spread out over a larger base.* This will help reduce stress, especially when lifting heavy objects or using heavy tools. For example, having enough personnel to affect patient transfers will generally low ergonomics risks.

● Use job rotation with caution and only as a preventive measure, not as a response to symptoms. Job rotation at an emergency scene is particularly important to reduce the potential for heat stress.

Work Practice Controls

An effective program for hazard prevention and control will include procedures for safe work practices; it must be understood and followed by all officers, staff, and line personnel. Key elements of a good work practice program includes:

- proper work techniques,
- fire fighter and EMT training,
- personnel body conditioning,
- regular monitoring,
- feedback,
- adjustments,
- modification, and
- maintenance.

Personnel must be trained in work practice controls that consist of demonstrations and practice time. The following examples show work practice controls:

- Task approach to improve posture and reduce stress on extremities;

- Proper lifting techniques;

- Proper use and maintenance of pneumatic and power tools;

- Use of ergonomically designed equipment and fixtures; and

- Proper use of PPE and other protective equipment.

All personnel must receive regular feedback on work practice controls. The following questions can be asked to ensure proper feedback:

- Are specified procedures being used?

- Are the procedures an improvement over the old method?

- Is re-training necessary?

- Are maintenance programs satisfactory?

Focusing on only one aspect of the job often will not prevent injury. Integrating multiple approaches can eliminate or control risk factors. To reduce CTD and other ergonomically-based injuries, an evaluation of the complete work environment must be conducted (see **Chapter 4**). Variables that must be addressed are job tasks, tools, and workplace design. Personnel should be trained to use proper methods, properly use specific tools or equipment, and recognize workplace design features.

Workplace/Workstation Design

Workplaces tend to refer to the total environment surrounding where the job/task is performed. A workstation is more related to a specific fixed place, fixture, or piece of equipment, and therefore more typical of manufacturing industry. The prevention or elimination of ergonomic injuries requires an evaluation of the total work environment. Workplace designs should reduce extreme and awkward body postures and movements. The locations of parts, tools, and tasks are best addressed during the design stage. Environmental factors such as heat, cold, noise, vibration, inadequate lighting, and poorly designed equipment can contribute to ergonomic disorders and should be considered during planning.

To assess any workplace, first determine what motions the task involves; then measure existing reach and height dimensions and compare these with guidelines. Workplace designs should be based on body dimensions, because people vary in size. The interrelationships between person, task work surface, seat, reach, and controls will differ from person to person and from task to task.

1. The height of seated or standing workstations should be adjustable for operators of varying size and shape. Materials and controls should be located so that hands do not strike objects during rapid movements. Edges, corners, and protrusions from surfaces should be rounded so that the individual can use them without rubbing their arms on the edge of the work surface does not cause injury. Positioning devices, should be used where possible to eliminate the need for the fire fighter or EMT to hold the work object.

2. Chairs must be comfortable and adjustable to meet the needs of a specific task. They must have an adjustable seat pan and backrest, and always be stable on the floor. In addition, the seat pan should not have an edge that causes pressure to the back or legs (water fall design).

3. Adjustments, such as platforms, should be sufficient to accommodate a variety of personnel and adapt to the physical characteristics of the worker population. Elbow, wrist, arm, foot, and backrests should be provided where needed to avoid static muscular work.

4. Equipment operators should be provided with floating seat belts and a passive suspension, adjustable to the size and weight. Active suspension seats, or seats that sense a motion and compensate with a counter motion, should be avoided.

5. Storage containers, especially if heavy, should be kept where personnel can reach them without extreme torso bending/rotation or shoulder flexion. Storage containers should not have hard or sharp edges. They should be designed to be held with the entire hand rather than in a pinch grasp with the fingertips,

Anthropometry.   Anthropometry can be described as the study of body measurements and the physical capabilities of the personnel. Anthropometric measurements are used to adapt for varying body dimensions and capabilities so that all personnel can perform the task.

Designing a workplace for the average male would not adapt to many individuals. The ergonomist's role is to design the workplace for approximately 95 percent of the fire fighter (or EMT) population.[6] Examples of this approach would be providing an adjustable-height workstation of having adjustable platforms available for shorter employees. Designing for 100 percent of the population is usually not economically feasible or practical. However, if a department has individuals who are either very large or very small, specific accommodations should be made.

Anthropometry studies the relationship of the physical features of the body to its environment. Weight, size, range of movement, and the linear dimensions of the body are all taken into account. Recommended dimensions for the workplace or suggested ranges for weight and force are derived through anthropometric analysis.

Once the anthropometric data are generated, they must be used correctly. The department must consider both the specific personnel and data's source. For example, placement of ladders on apparatus for effective work heights for removal and storage based on a predominantly male work force will not probably account for the shorter, less-physically fit persons.

Some examples of other workplace controls include:

Proper Illumination. Detailed work makes great demands on the eyes. If the illumination is inadequate and not glare-free, the worker must lean forward or to one side to see clearly. Eyeglasses and other eyewear must be correctly fitted. Personal protective equipment must accommodate eyeglasses or provide for the attachment of corrective lenses.

Work Surface Height. Usually adjustable work surface provide a better work position. Like the adjustable height mechanisms, the tilting mechanism should be simple, and preferably have a nonskid surface to prevent parts from sliding. There is often confusion about the proper height 'work surface that should be purchased. In the fire and emergency medical services, it is especially important that the height accommodate the various postures and motions of the individuals. Common examples of work surfaces include patient stretchers. A workbench that is too low would force individuals to bend their back or neck, and rotate their shoulder forward. A workbench that is too high can force employees to work with elevated arms. Both situations can lead to rapid fatigue, pain, and before long, injury. The height of work itself is usually more crucial than the height of the workbench. Often a compromise must be made for the comfort of the arms, back, and neck. Because it is best not to overstress particular body parts, the workplace height should equally distribute the stress among all body parts. This will reduce the potential for injury. However, the nature of the work also affects the height level. For example, heavier work should be lower and lighter work should be higher. There also may be mismatches in heights between various pieces of equipment, causing unnecessary motion and exertion.

---

[6]Volume 1 of **Ergonomic Design for People at Work** (Eastman Kodak Co., New York, Van Nostrand, 1983) provides detailed information on the application of anthropometric data for general populations.

Step Height. Numerous fire fighter injuries occur getting onto and off of apparatus and other emergency vehicles. Proper placement of steps combined with appropriate handhold and adequate illumination of offloading areas is important to eliminate this types of injuries. Also related are station apparatus bay surface non-skid finishes which can prevent slipping and the associated acute trauma.

Displays. Many fire fighter and EMT tasks require information processing. It is particularly important for some personnel; for example, dispatchers, company officers, and apparatus drivers. Clear communication of information depends on good sensory input, adequate lighting, and lack of noise. Complex information should be visual. Simple or urgent information can be auditory. Combining both is best; the auditory signal alerts the operator, whereas the visual display adds detail. For inspection purpose, complex displays and perceptual organization can be an important factor. People have a limited capacity to perceive information. Grouping and simplifying information reduces demands and makes errors less likely.

Job Tasks

In order to modify the task, the physical aspects of the job must be analyzed. One of the best ways to reduce lifting injuries is to avoid placing objects in the worker danger zones. One of the danger zones is below the person's knees. Picking up or setting down an object lower than the knee amplifies a risk for back injuries. Instead of placing objects on the floor, place them on tables or other surfaces elevated at least to knee level. If the objects must be stacked so that the level of the pile is constantly changing, an adjustable height platform could be used.

The other danger zone is above the person's shoulders. The shoulder is rather poorly designed, biomechanically speaking, to withstand extreme forces in the overhead position. Therefore, in jobs involving significant weight or repeated lifting, work performed above the shoulders should be avoided. Removal of fire hose from the top of the apparatus is an example of an operation which could cause injury. However, proper placement of the hose can reduce the number of fire fighter over head reaches and activity.

Another cardinal rule is to avoid severe rotation or twisting of the spine when lifting. When personnel hold a load in the hand, most of the force of the load (and the weight of the upper body) is applied to one spot in the spinal column. The weight of the load and the twisting can provide more stress than the vertebrae and the disc can take; slipped, ruptured, or herniated discs can result.

In one hand lifting, pushing, and pulling, there is also danger of the spinal column being pulled to one side. The vertebrae of the spinal column contain discs between the bones to provide cushioning and to allow the bones to move more easily. With excessive pressure on one side of the spinal column, it is possible for the disc to be squeezed to one side. This can result in a herniated or slipped disc, both are extremely painful and expensive conditions to treat.

Therefore, it is important to work both sides of the body equally. There are several ways to accomplish this; one is to move the bin to the other side for part of the day. The U.S. Postal Service worked out a similar solution for mail carriers found to carry the sacks on their right shoulders during the day. When the mail carriers switched shoulders during part of the day, their symptoms decreased. When possible, lifting load should be balanced.

It is possible to avoid lifting through the use of mechanically assisted equipment. Hoists have been used for handling heavy loads; unfortunately, they can slow responses--so fire fighters sometimes don't use them. When developing ergonomic controls, avoid procedures or equipment that will slow response time. Under critical emergency conditions, if lifting cannot be avoided, there are ways to make the mechanics of the lift a little easier. One way is to use a padded rail that the employee can lean against when bending to perform a lift. The height of the rail may be slightly below the waist or at mid-thigh level, This gives the person a little extra support that reduces some strain on the back. A similar device is available; it is a framework that supports the body as it bends and performs other tasks.

Reducing Fatigue and Awkward Postures

There are many ways to reduce the fatigue factors. The following list outlines methods that can help reduce fatigue:

- reducing force, duration, and frequency of exertions,

- eliminating unnecessary movements when transporting materials,

- supplying material or tools in a defined sequence and in the proper relation to the work,

- avoiding poorly fitting clothing and gloves that require forceful exertions to manipulate,

- providing posture changes between standing and sitting,

- providing evenly spaced work breaks, and

- avoid heat stress by replacing electrolytes.

When possible, task should be undertaken so personnel do not have to work with body parts in an awkward position. Exertions in awkward positions may require muscles to work at or near maximum capacity. Muscles cannot maintain static contractions greater than 15 to 20 percent of their strength without fatiguing. Therefore, tasks should be designed to reduce these awkward positions. When evaluating or designing a work area, be sure to review the frequency of the following actions:

- the wrist not in the neutral position (not ulnar or radially deviated),
- elbows not close to the body and bent 90 to 110 degrees,
- number of reaches over shoulder height,
- number of reaches behind the body,
- lateral bending requirements,
- twisting or bending the torso requirements,
- requirements for bending the neck backward requirements,

- bending the neck forward (more than 20 to 30 degrees),
- wrong height in relation to tasks, and
- requirements for kneeling.

Seating

Seating should be viewed in relation to a specific task. It is usually a dynamic not a static activity. People need to be able to move around in their seats, lean to retrieve material, get up, and sit down. For many tasks, individuals are seated at something: a work surface, a bench, a console, etc.; therefore, the seat should be seen in relation to the whole workstation. Because people vary in size, seat pan height and depth, arm rest, and backrest height should be easily adjustable.

Adjustable Chairs. The height of the seat is very important since some tasks require the worker to sit most of the day. The height of the chair should be easily adjustable within the range of 15 to 22 inches. The seat back should also be adjustable, up and down, and forward and back. Arm supports that are too high can cause more problems than it resolves. An alternative is a work surface with a semicircular indentation by the seat so that the arms can rest directly on the table surface.

Seat Area. A small weight-bearing area will result in increased pressure and restricted blood flow. This can cause considerable discomfort in the buttocks and thighs. Edges and ridges on the seat can cause similar problems especially at the front edge. Sweat can also cause discomfort if large areas of the thighs are kept in prolonged contact with the seat. Impermeable seat materials will worsen this situation.

Seat Depth. A seat pan that is too deep can result in painful pressure behind the knee. This pain encourages slouching and, therefore, loss of thigh and back support. The front lip of the lip of the seat should be rounded.

Seat Height. If the seat is too high, legs will dangle, increasing pressure on the underside of the thighs. If the seat is too low, knees will be raised, putting leg muscles under tension and creating leg-space problems. It is preferable to give footrests to shorter people rather than to restrict leg-space for taller people.

Back Supports. Back support is important and should help to maintain the natural vertical curvature of the lower spine and the horizontal curvature of the back. Back supports should be adjustable both horizontally and vertically. Shallow back supports cut into the back while very wide back supports may prevent the elbow from being drawn back,

Arm Supports. Arm supports are often necessary in industrial situations to prevent the arms from being supported by the work surface. The arm rests may prevent the chair from being drawn up to the work surface. If used, the arm rests should be far enough apart to accommodate broad people and wide enough to support the whole forearm. They should be adjustable in height so that slouching or hunched shoulders can be avoided.

Controls. Communication between people and machines can be made more efficient and less stressful by taking a few principles of control design into account. Once sensory information has been processed and a decision has been made, some action will be necessary. This action is usually achieved by operating a control. Operators should be able to find their controls easily while viewing the displays or the operation itself. Make controls easy to locate by appropriate spacing, and easy to identify by touch. When the body is tired, under stress, or inexperienced, it tends to act instinctively. If control operation contradicts common habits or is confusing, errors are more likely to occur. For example, in the U.S., we typically turn switches up or right for ON. For the same reason, the display response must match the control action. If the control moves up, the display should move up. For vehicle controls, make the control match the action it is controlling. Standardization must play an important part.

Back Belt Use

The use of back belts for injury prevention is a controversial issue. Proponents advocate use of an elastic-type back support belt for use on-the-job by healthy, previously uninjured workers. Other groups suggest that the efficacy of such belts remains unproven and that injury rates may in fact increase, due to restricted mobility or acquired dependency.

The controversies arise over back support belts designed for general population use. These are to be differentiated from therapeutic devices, including spinal braces, supports, corsets, and orthoses, that may be prescribed by a medical professional for treatment and rehabilitation of injured workers. In addition, the leather weight-lifter's belt used by athletes have not been recommended for general use.

Advocates of preventive use of back belts claim that the belts:

- reduce pressure on the spine ("loading") caused by lifting;
- increase intra-abdominal pressure (IAP), thus counterbalancing the compressive force on the spine;
- stiffen the spine, thus reducing disc damage by decreasing motion in the spinal segments;
- reduce bending motions;
- reduce certain workplace injuries; and
- remind workers to lift properly.

A 1994 NIOSH investigation evaluated various studies in which the above results were claimed. The NIOSH Working Group concluded that there is little or no evidence from these studies to support the claims. The Working Group did not recommend the use of back belts to prevent injury among uninjured workers, did not consider back belts to be protective clothing, and stated that back belts did not mitigate the hazards of repeated lifting, twisting, pulling, pushing, or bending.

The NIOSH publication, "Workplace Use of Back Belts," describes the Working Group's evaluation of the various studies and its recommendations for further studies. An extensive bibliography of back belt studies is included. "Back Belts - Do They Prevent Injury?" is a pamphlet that summarizes the NIOSH recommendations, provides answers to general questions about back belts, and emphasizes the need for an overall ergonomics program to reduce injuries.

Sources for NIOSH publications are listed in the box below.

PRINCIPAL REFERENCES RELATED TO BACK BELTS

Publications

1. Waters et al., "Revised NIOSH equation for the design and evaluation of manual lifting tasks," *Ergonomics, 1993, vol. 36,* no. 7, pp. 749-776.
2. *Lift Guide*^{TM}*computer software for the revised NJOSH lifting equation,* National Technical Information Service (NTIS), 703-487-4650.
3. *Applications Manual for the Revised NJOSH Lifting Equation,* Order number PB94-176930LJM, (NTIS), 713-487-4650.
4. "Back Belts - Do They Prevent Injury?," DHHS {NIOSH) Publication No. 94-I 27, contact NIOSH at 800-35-NIOSH.
5. "Workplace Use of Back Belts," DHHS (NIOSH) Publication No. 94-122, contact NIOSH at 800-35-NIOSH.

Training Programs

1. Back Power, National Safety Council, 800-621-6244 (back injury prevention program emphasizing improving back fitness).
2. Back Safety - The Ergonomic Connection, Summit Training Source, Inc., 800-842-0466 (program addresses body mechanics and preventive measures).
3. Back Basics and Protecting Your Back, NUS Training Corporation, 800-338-I 505 (videos addressing back exercises and lifting techniques).

NOTE: There are other useful references and training programs on back belts not cited here. The U.S. Fire Administration and the Federal Emergency Management Agency does not endorse the above companies.

Whether or not back support belts are used, all groups agree on the importance of an overall ergonomics program to reduce back injuries. In addition to considering job or equipment redesign, education on proper lifting techniques and back fitness is paramount. Resources for lifting or back safety programs include programs such as "Backpower," designed by the National Safety Council. Representative sources for a variety of back training program are listed in the box below.

Fire service members who have used back belts have varied opinions on their effectiveness. Many departments issue them on request for voluntary use. Actual usage figures are not available. Many users believed that the major benefit is that the belts might serve as a prompt to remind the wearer to use proper lifting techniques. Some anecdotal reports suggested a possible increase in injuries caused by restricted mobility.

If the decision is made to use back support belts, the following factors should be considered:

1. determination of activities during which back belts would be worn and whether compliance is voluntary or mandatory;

2. compatibility with department's fire fighter protective clothing;

3. fire resistant properties of belt material; and

4. belt design (to insure required mobility and desired wear properties). Proponents recommend "spring" steel stays (at least 0.014" thick) with heavy plastic coating on ends, incorporated into a belt of two-way stretch elastic (spandex "open weave" power knit) with two-way stretch elastic binding.

Several companies have introduced turnout gear with integral lumbar support systems. These system designs eliminate dependence issues such as muscle atrophy because the support is used only during responses. However, because these products are relatively new, long term efficacy studies have not been performed. These types of back support were not evaluated in the NIOSH study cited above. The manufacturers should be contacted for up-to-date data on these products. In addition, other companies may introduce related products.

1. One manufacturer offers a turnout coat with removable lumbar support. Globe supplies a turnout pant with a detachable lumbar support that is activated by an attached self-contained aspirator. Activating these supports is more easily accomplished with the coat unfastened; however, deactivating either system could be performed easily with the coat fastened. If desired, the supports can be worn independent of the bunker gear for EMS calls.

2. A second manufacturer has recently introduced a fire fighter pant incorporating a lumbar support system. The system can be activated or deactivated externally by means of hook and loop type closures. Field tests and anecdotal reports from fire fighters are favorable. EMS jumpsuits and pants are available with the support system.

The International Association of Fire Fighters has published an excellent guidance summary regarding back belts and sends out a collection of articles supporting their recommendations against the use of back belts, in general, and prescription of exercise as a prevention strategy.[7]

Ergonomic Principles of Hand Tool Use

Ergonomically engineered tools have been around for 30 years and are only currently

---

[7]Back Injuries and the Fire Fighter, IAFF Publication, Department of Occupational Health and Safety, 1750 New York Avenue, Washington, DC, Revised 6/27/95.

beginning to show up on the market. Nevertheless, injuries are still occurring from repetitive use of hand tools, particularly when awkward hand positions and/or forceful exertions are involved. Even with this new technology, there are still many manual cutting operations. Various injuries can occur where there is still use of such tools.

Fire fighters and EMTs use a variety of tools and equipment in the responses activities. Examples include:

- forcible entry tools or irons,
- spreaders or wedges,
- rescue and bolt cutters,
- pike axes and picks,
- mallets and mauls,
- saws,
- hydraulic lifts, and
- flash lights.

Ergonomic principles should be kept in mind when ordering tools or purchasing new equipment. A review of the ergonomic guidelines will help the purchasing agent understand these principles. Ergonomically correct tools may cost more as an initial investment. However, the long-term savings in preventing illness and injury will more than pay for the extra cost. Considerations for selection include the following:

1. Before instructing a fire fighter or EMT to use new equipment, it is important to *find out how the equipment feels when begin used.* Tools should be tried out and the potential users should understand how they are used and how they feel before being specified.

2. *Hand tools should be selected which are designed so that the wrist can maintain a straight and neutral position.* Whenever possible, tools should be chosen that have been designed to reduce vibration and which do not transmit torque to the hand and arm. Power tools are preferred because they reduce the force and exertion required to perform the task. All devices must be designed for safe operation.

3. *Tool balancers should be used where possible.* The center of gravity must be aligned with the center of the grasping hand; this provides the leverage necessary to keep the tool in alignment. Tools that are continuously held should be weigh approximately one to two pounds. Any heavier tools should be counterbalanced.

4. *Handles should contact as much of the hand and fingers as possible,* with a diameter approximately 1.25 - 1.75 inches. The minimum handle length should be five inches. Tools requiring both hands should provide two handles. Handles should be positioned to reduce awkward positions, and have a maximum distance of two - 2.7 inches. Tools should be equipped with a comfortable grip span between the thumb and forefinger. Vibration dampening materials should be

incorporated in or on tool handles. These materials can also be incorporated into gloves. Where possible, use slip-resistance material for tool handles.

5. ***Narrow tool handles which concentrate large forces onto small areas of the hand or grips with finger grooves, ridges, or recesses should be avoided.*** Short tool handles that press into the palm of the hand and tools that exert force onto the sides or back of the hand. Power tools that must be grasped by the motor housing should be avoided. Evaluate pinch points, sharp edges, vibration transmitted to the user, tool exhaust air directed toward the wrist or other parts of the body, and excessive noise.

6. ***Safety features should be maintained as originally designed.*** Knives and other tools with a cutting edge must be keep sharp. A regularly scheduled tool maintenance program should be established.

One tool maker's guide states that "a short term high load at the limit of capacity is generally preferable to a long term low load at the limit of capacity." Manufacturers therefore attempt to make the tool as powerful as possible within a given weight and volume. This may lead to high momentary surface pressure between the hand and the tool. This condition places particular demands upon tool design, but is less demanding physically from a human point of view than prolonged load exposure.

Hand tools must permit tasks to be performed with minimum force and allow the body joints to be as close as possible to a natural position. When selecting tools, it is important to consider characteristics such as weight, shape, balance, and vibration. Weight and vibration can stress body joints and tissues. Tool shapes can determine the position and motions required by the worker to complete the task.

Powered and Air Powered Tools

Powered hand-held tools are used in many different types of industries. The combination of precision workmanship and high power tools enable workers to carry out complicated tasks faster than using a manual tools. A well-designed workplace equipped with carefully chosen power tools often results in a higher productivity.

There are many ergonomic advantages to using air-powered tools instead of manual tools, Forceful exertion with a hand tool can create many CTD problems. A manual tool requires significant exertion which can put the employee at risk of injury; with air-powered tools, the only exertion would be the least pressure of the thumb or finger to activate the trigger. On many power tools even this is not required since an automatic sensing device or foot pedal can activate the tool.

Air-powered tools eliminate the excessive forces that are necessary when manually cutting through heavy material. This significantly reduces the risk of injury, while it increases productivity. Operating manual cutters involves the exertion by the fingers and thumb. This involves continuous flexing of the finger. In contrast, most air-powered cutters are activated with the thumb. This eliminates the use of the finger and relies instead on the stronger thumb muscle, Many manual cutters and shears have such large grip spans that employees are forced to hold and operate the tool with their fingertips.

It is best to avoid the pinch grip, which is weaker and makes the hand prone to injury. Manual tools require the employee to use a pinch grip (the thumb opposes fingers, although it is a wide span). In contrast, air-powered tools allow the employee to use the safe, whole-hand power grip. When switching to air-powered tools, the time needed to perform the task should be significantly reduced, by that increasing productivity.

Considerations in Selecting Hand and Power Tools

Except for vibration and switch actuation methods, the problems associated with non-powered hand tools are similar to those of powered tools. These problems include awkward hand positions, stress on tissues or joints, excessive force requirements, and handles that are not appropriate for the user's hand.

- Tool handles should be long enough to extend past the user's palm. If the worker is wearing gloves, another half-inch of length is needed. Their surface should be broad enough to distribute pressure evenly, and they should be padded or slip-resistant.

- Tool handles should not have sharp corners or edges. Handles like those on pliers or scissors should be designed to open with a spring or other device. This avoids the exertion of force with the side or back of the hand.

- The muscles of the hand function best in the middle of their range of motion, when they are neither fully flexed nor fully extended. Handles like those on pliers, wire strippers or scissors, should be keep in this middle range.

Vibration. Some workers who are continuously exposed to the vibration of power tools may experience circulatory problems. The length of exposure to the vibration is a major factor, as is the frequency. If possible, frequencies between 40 and 90 Hz should be avoided, although low-frequency vibration are also harmful.

Activation Method. It's a good idea to find an alternative to a power tool that is started and stopped by a one-finger device. This device can be replaced with: a lever arm that can be manipulated by several fingers; air-operated start methods; push-start methods on the drive shaft; or a switch operated by the thumb, rather than a finger. A "dead man's" switch may be of value as an injury prevention device with many power tools.

Torque. Driving a screw with a power tool is likely to transfer torque to the worker's hand when the screw bottoms out. This is experienced as a snapping action, and the repeated stress is potentially a serious problem. This stress can be reduced by using slip clutches or torque limiters, keeping the torque settings low, or mounting the tool on an articulating arm that keeps the torque from reaching the hand. Another solution is to provide an extra handle so the worker can use two hands to help counter the torque effect.

Force Requirements. The force required to perform a task is related to the posture of the hand, the characteristics of the object grasped, and the friction between the hand and the object. Force can be reduced in a variety of ways. The tool handle can be altered to make it more efficient, or the weight of the object can be reduced. The task can also be altered to require less exertion with each action. Tools should be designed to be held at their center of gravity, so their weight cannot twist them out of the hand. When the hand location cannot be easily changed, change the center of gravity by reducing the weight of the tool, shifting its weight, or adding weight to the light end. Force requirements are typically greater for non-powered tools than for power tools performing the same job.

Hand and Wrist Positions. As the hand moves away from its natural posture, stress increases on the nerves and tendons. Awkward wrist positions can be caused by the nature of the tool handle or the position of the object. Several manufacturers offer tools that can improve the situation in certain awkward tasks. These tools are usually bent to allow the tool to assume the awkward position, instead of the hand. However, these tools often have a limited range of uses. It's often necessary to modify the tools in the workplace to achieve a specific result.

Pistol Handles. The pistol handle is used to shorten the length of the tool and reduce the bending load on the wrist. Keeping the distance between hand and work piece as short as possible allows precision tasks to be performed efficiently. For tasks requiring high feed forces, the hand needs to be angled at 70 degrees. Where only twisting forces occur, the center of the tool's weight should be on top of the hands. This keeps the bending force on the wrist low. One handle has been developed to apply high feed forces and resist twisting forces. The high grip keeps the straight line through the arm and wrist for high feed forces. The low grip transmits twisting forces comfortably to the wrist.

Straight Handles. When held, the diameter of handles should allow the strongest grip, and cause minimum strain on the hand. The next aspect to consider is whether the tool produces a twisting torque in the hand (screwdrivers and drills) or produces a levering torque on the wrist and arm (grinders and angle tools).

Triggers. The trigger is closely related to the design and function of the handle. This is used mainly on pistol handled tools where the trigger can be operated without affecting the grip on the 'handle. The trigger should be designed to ensure that it is pressed with the middle part of the finger. If the tip of the finger is used, tendon can result. This type of trigger gives precise control over the speed of the tool. To give the worker the right feel, the trigger should be designed so that to increase speed, the pressure on the trigger must be increased. Finger triggers are used for frequent operations, precision tasks, and jobs where the tool needs to be positioned before it is started.

Strip Triggers. With strip triggers, operation is achieved with low trigger forces. All fingers are used to minimize strain in the hand. With this type of trigger, the grip on the tool is unstable. With the lever start, the trigger operation becomes part of the grip on the handle, giving greater stability, ideal for long cycle operations and for tasks which place high forces on the tool. It enables the tool to be positioned before the start-up; in addition, a safety device can be built into the trigger without affecting its operation. The lever start triggers are used on grinders, drills, and screwdrivers.

Thumb Triggers. This is a trigger that can be operated while maintaining a grip on the tool. This is suitable for tools, such as chipping hammers, because high feed forces are used. In this application, the trigger operation becomes part of the feed force.

Push Start. Push start is used primarily on straight screwdrivers because it is ideal for repetitive high volume operations. The trigger is activated by putting a feed force on the tool. A stable grip is always maintained on the handle.

Stall Tools. With this type of tool, the fastener is directly connected to the motor through a gear train. All the torque applied to the fastener must be reached by the operator who is holding the handle of the tool. For certain applications, a ratchet clutch can be put into the drive line; this does not affect the torque reaction. These tools tend to be large and heavy.

Impact Wrenches. Impact wrenches contain a mechanism that consists of a rotary hammer and an anvil, which contains the fastener. During run-down, the hammer and anvil rotate as one. When resistance is met, the anvil is stopped. However, the rotary hammer continues to be turned by the air motor. More torque is applied to the fastener when the rotary hammer hits the anvil. The only torque reaction occurring in the handle is that caused when accelerating the hammer.

Shut Off Tools. These tools have a clutch mechanism that shuts off the air supply to the motor at a pre-set torque level, which is built into the drive line. With a fast-acting clutch, the inertia in the tool takes most the reaction force (on most joint conditions). These tools are not as compact as the impact/impulse tools, but are small and fast compared to a stall tool.

Tool Balancers. Where possible tools should be attached balancers. Balancers reduce the strain on the worker, prevent damage to the tool and keep the work area tidy. There are two types to choose between: the standard spring balancer and the constant tension balancer. The standard spring balancer applies a force that retracts the tool to a stand-by position. This force increases as the wire of the balancer is further extended. Every time the tool is used, the worker has to overcome this load. The constant tension balancer applies a constant force on the tool despite wire extension. This reduces the load on the worker.

Apparatus Design

Because of the emergency nature of most activity related to fire apparatus, and because much of it tends to be rather strenuous (heavy pushing, pulling, and lifting), there are phenomenal demands made on fire fighter footing. It is therefore not surprising that many serious injuries arise from slips and falls off of fire trucks, since it is not unusual for such equipment to respond to several calls each day. The following guidelines should be used in the design and specification of apparatus:

- Use dimple plate construction on step surfaces which is slip resistant, self-cleaning, and self-sharpening;

- Keep apparatus access ways uncluttered;

- Avoid steps which are too small;

- Provide slip-resistant ladder cleats;

- Attach sturdy grab bars on the sides of large step plate;

- Use a maximum safe distance of 24 inches from the ground to the first step;

- Specify grab bars which are non-round to promote better grip;

- Employ retaining pin chains designed to prevent breakage;

- Locate accesses to storage spaces such as handles or levers closer to the bottom of doors;

- Provide ground lights below the running board, front bumper, and rear step to aid working and moving around the vehicle;

- Incorporate slide-out shelves in storage space to bring heavy objects out away from the apparatus to make lifting easier; and

- Avoid placement of heavy boxes and equipment in high storage spaces.

Since it is difficult to correct problems with apparatus design after purchase, it is important that department representatives seriously consider ergonomic concerns in the specification or evaluation process.

Use of Personnel Protective Equipment

Personnel protective equipment must be used by fire fighters and EMTs to protect them from the various hazards of the workplace, such as heat, flame, and blood borne pathogen exposure. Nevertheless, this same clothing is generally encumbering, limits mobility, and can create ergonomic stress. It is important that clothing be properly designed and use much in the same way as tools to achieve the maximum benefits of its protective qualities without a significant decrement of the wearers performance. Most standards on protective clothing do not address human factors performance. Therefore, these issues must be ascertained through field studies or trial evaluations of clothing and equipment items.

When possible PPE should be:

- as light as possible,

- not restrict fire fighter (or EMT) movement,

- be appropriate sized (and available in sufficient sizing for correct fit of department personnel, and

- function as claimed by the manufacturer.

Since 'breathable' protective clothing is known to provide lower stress on individuals during light and moderate work loads as demonstrated in a number of field studies, protective clothing with 'breathable' moisture barriers should be selected. However, protective clothing that is light and breathable cannot be depended on for reducing stress. Reduction of clothing stress must also be managed through providing appropriate fire ground staffing (to reduce and distribute individual workloads), work-rest cycles where they can be implemented, and rehabilitation after heavy work.

Improper footwear and lack of ankle support are major causes of ankle and lower extremity injuries. Footwear which provides good fit to the individual, especially in terms of ankle support, and having relatively wide soles should be selected over poorly designed, inadequately designed footwear which does not provide normal fit and ankle support.

Other Specific Equipment and Work Considerations

Several other areas presenting specific ergonomic hazards to fire fighting and emergency medical operations can be addressed by better equipment specifications or work practice controls. Examples of these design/work practice solutions are presented in Table 6-1.

CHAPTER 6 - TRAINING

Most fire and emergency medical services departments realize the benefits of continual training. Training helps to reinforce new concepts and keep awareness levels at a minimum level for all employees. These are particularly important goals when a new program is implemented within the department. In training to support the implementation of an ergonomics program, several questions may arise:

1.   What are the department's or organization's training needs?

2.   What are the goals and objectives of the training?

3.   What should the content of meet these department training goals and objectives?

4.   Who should conduct the training and how frequently should the training be accomplished?

5.   How should the effectiveness of the training be judged?

The Importance of Training

Training is an important tool because it enables the fire or EMS department to acquire the practice expertise for implementing ergonomics in a systematic way that will have results. *Training provides a proactive approach to solving ergonomic problems because fire fighters and EMS personnel are equipped not only with the methods to solve existing problems, but to make decisions that will prevent problems in the future.* For example, fire fighters can adjust and rearrange equipment on the apparatus to make it easier to reach and retrieve. They can also make the decision that any future apparatus designs will include specifications which address potential ergonomic hazards.

Training is also important because:

● It allows the employees to participate in resolving ergonomic problems.

● It can motivate them to take actions that contribute to their own well-being and to suggest ways of improving their existing jobs.

● Participants learn when they need the assistance of an expert to help in finding solutions.

Identifying Training Needs

Training needs can be identified by examining the problems in the specific department, station, or unit and determining whether they can be solved by training. Problems that can be effectively corrected by training are those that arise from:

- a lack of knowledge of a response or work task,

- unfamiliarity with equipment,

- incorrect performance of a task, and

- exposure to new or different conditions at the emergency scene.

Not all work-related problems of the fire and emergency medical services can be corrected by training in ergonomics, especially since the nature of emergency responses means that many conditions at the emergency scene cannot be anticipated or provided for. However, many tasks have several functions in common which are can affected by ergonomic-based controls.

If training is deemed necessary, the next step is to review the department's ergonomic job hazard assessment **(Chapter 4)** to identify which personnel are subjected to specific risks. Examine injury records to determine which areas deserve the most attention. This will help design the training content of the program. Let field personnel be a source of valuable information to help you in identifying safety and health hazards.

Training Goals and Objectives

Once the department's training needs have been identified, clear measurable objectives should be developed. These objectives should:

- include what department personnel are to learn and how they will demonstrate what they have learned.

- allow definition of competent and acceptable job performance in terms of ergonomic principles. For example, fire fighters should learn to only lift extremely heavy objects with the assistance of another fire fighter.

Learning exercises must be matched with the training program objectives. To ensure that department personnel are able to use the skills or knowledge from the training, the learning exercise should simulate the actual job as closely as possible. Whatever methodology is used, lectures, group discussions, drills, simulations, etc., it is important that the training program allows personnel to demonstrate clearly that they have acquired the desired skills or knowledge.

Content of the Training Program

Ergonomic training program content must reflect the objectives and goals set forth by the department. In general, this training must be accomplished be accomplished for recruits and then repeated throughout the career or involvement of response personnel. In many cases, ergonomic training can be effectively accomplished as part of the department's overall safety and health training program as defined in NFPA 1500.

Although many people think of general fire fighter or EMT education when they hear the word "training", the scope needs to extend to the entire organization, addressing the issues that are most appropriate to the various roles of the participants: department chiefs, division managers and their staff, and support personnel. Different levels and functions of personnel within the department need different amounts and types of training as shown in Table 6-1 below.

Table 6-1. Training Levels by Department Personnel

Personnel Level	Length of Course	Course Focus
Department senior officers and department or division managers	1 to 2 hour session	strategic importance of ergonomics in the organization for reduction of workers' compensation costs and increase response efficiency
Safety officers and ergonomic committee members	2 to 3 days	formulation and implementation of written ergonomic plan that defines a systematic approach to ergonomics; recognition of ergonomics hazards and ergonomic disorder symptoms; design of solutions to ergonomic programs
Equipment specifiers and purchasers	varies	specific ergonomic concerns related selection of tools, equipment, and apparatus
Station leaders and captains	4 to 8 hour session	overview of ergonomic principles; ergonomic safety and health concerns regarding the employees; support for implementation of ergonomic program
Fire fighters and EMTs	2 to 4 hour session	ergonomic principles and the practical applications that they can make; what are and how to report symptoms of discomfort early so that solutions can be found before severe conditions develop

All classroom training, even though tailored for the specific audience (e.g., management versus line personnel), should include the following minimum material:

- general principles of ergonomics,
- types of CTDs and other ergonomic disorders,
- risk factors contributing to ergonomic disorders,
- recognition and reporting of symptoms,
- methods of preventing ergonomic disorders,
- basic human anatomy and physiology.

The outline for an example training course content for general fire fighters and emergency medical technicians is given on the next page.

EXAMPLE TRAINING COURSE OUTLINE
FOR GENERAL FIRE AND EMS PERSONNEL

1. Introduction to Ergonomics
   a. Definition of ergonomics and why understanding ergonomics is important for safer and more efficient responses
   b. Department policy and program involving ergonomics

2. Basic Principles of Ergonomics
   a. Types, recognition and avoidance of ergonomic hazards (hazard identification)
   b. Basic human anatomy and physiology
   c. Cumulative Trauma Disorders (CTDs)
   d. Other ergonomically-related muscular-skeletal disorders
   e. Heat stress, cold stress, noise stress, visual stress and other response scene problems

3. Prevention and Control of Ergonomic Hazards
   a. Physical training to prevent injuries
   b. Proper lifting technique
   c. Suggested techniques for other difficult or awkward tasks
   d. Reducing impact of environmental stresses
      Tool selection and use
   f. Apparatus safety
   g. Use of PPE

   HANDS-ON EXERCISES

4. Dealing With Injuries
   a. Early recognition of symptoms
   b. Reporting and recording injuries
   c. Department policies regarding light (alternate) duty, rehabilitation, and workman's compensation

5. Review of key concepts

   EXAMINATION

   COURSE CRITIQUE

Methods for Conducting Training

Training can be accomplished in a variety of different ways. Training may be conducted by immediate supervisors, safety department personnel, or outside experts. There should be some level of ergonomics expertise in each department so that reliance on outside sources can be avoided. However, if you choose to use outside experts, the following characteristics of effective training should be used:

1. The training needs to be highly interactive, consisting not only of a lecture, but hands-on activities, slides, and videos of the particular work environment.

2. The training material should be totally integrated as a training manual or package, not just a collection of articles and papers.

3. The person conducting the training needs to have the appropriate credentials for the audience.

The training must be structured around specific objectives that will enable the participants to leave with an action plan in place so that implementation can begin. Adults learn best when the knowledge has relevance to their needs and is related to their experience. For example, fire fighters and EMS personnel are more likely to respond to someone from the fire or emergency medical services, or at least some one who can relate specific experiences in fire fighting or emergency medical operations, as opposed to individuals who typically train industrial/ manufacturing employees. A mixture of lecture, visual (films, videos, or overhead transparencies), discussion, and hands-on exercises is often the best approach.

As with any other type of safety training, it is important to tell the fire fighter/EMT why they are being trained. This helps them to associate the lesson with an understanding of the subject they are studying. Providing examples of how ergonomically-based injuries have resulted in prolonged treatment and lost time for fellow fire fighters and EMTs usually gets attention for justifying their involvement and participation in the program.

Good personal work habits can go a long way in preventing repetitive stress injuries. Training fire fighters and EMS personnel how to lift properly, use tools effectively, and avoid awkward motions should be required in recruit training programs. Supervisors and division managers should be repeatedly trained to recognize the symptoms of cumulative trauma syndrome and other ergonomic-based injuries. It is also important that all personnel learn to recognize the effects of environmental stresses (heat, cold, noise, visual) and understand the limitations of personnel protective equipment (PPE) when used to protect against these hazards.

Motivating personnel to learn and understand the material is an important part of accomplishing effective ergonomics training. The following steps can help to motivate employees and to clearly present the program:

● Explain the goals and objectives of the training (and how all department personnel will benefit).

- Provide an overview of the material to be learned.

- Relate the training to the personnel skills.

- Allow personnel to practice their newly acquired skills and knowledge to help correct any problems that may surface.

- Encourage personnel to ask questions, contribute their knowledge and expertise, and learn through hands-on experience.

Supervisors should also spend time explaining ergonomics to the fire fighter or EMT. This is especially important because some personnel, when left on their own, do not take full advantage of the knowledge for how to work without incurring injury, especially under the duress of an emergency.

Frequency of Conducting Training

The program content, the conditions at the response scene and workplace, and the training resources available will help you decide how frequently training should be conducted, the length of the sessions, and the instructional techniques to use. *At a minimum, all personnel should have some form of ergonomics training twice each year as recommended in NFPA 1500 (Paragraph 3-2.1).* At larger, busier departments it may be necessary to break training up into modules. Alternatively, it is possible to combine ergonomics training with other parts of the departments occupational safety and health training program or in specific response training; however, for recruit training, it is ideal to make ergonomics a separate module to provide emphasis to this area.

Evaluating the Training Program

To ensure that the training effort is meeting its objectives, a plan for evaluating the training sessions and methods of record keeping should be established when the course objectives and content are developed. Such evaluation can give you the information you need to determine whether your employees achieve the desired results and whether training should be offered again. The use of student test scores, supervisor's observations, and changes within the department which result in reduced ergonomically-based injuries and illnesses are tools you can use to evaluate the training program's effectiveness **(see Chapter 10).**

CHAPTER 7 - MEDICAL MANAGEMENT

Medical management focuses on early identification and treatment of ergonomically-based disorders, and prevention of future problems. Personnel should be encouraged to report early symptoms of these disorders, and should not fear reprisal or discrimination when they report symptoms. Many departments will already have the groundwork laid for such a program. Chapter 8 of NFPA 1500, *Fire Department Occupational Safety and Health Program*, specifies requirements for medical examinations, physical performance, physical fitness, fire department physicians, and post injury/illness rehabilitation. *Specific requirements for the ergonomic medical management element should be integrated with the medical and physical portions of the department's overall occupational safety and health program.*

This chapter is intended to describe the specific aspects of medical management as it related to ergonomics. There are two primary parts of the medical management element in the ergonomics program:

- The first part deals with **preventative measures** and includes baseline and follow-on medical physical examinations, recruit fitness training, regular fitness training, and annual fitness evaluations.

- The second part involves **treatment** and procedures following the diagnosis of an ergonomically-based disorder including long term actions such as rehabilitation and workers' compensation.

**ESSENTIAL PARTS OF THE MEDICAL MANAGEMENT
ELEMENT OF AN ERGONOMICS PROGRAM**

1. Preventative Measures
   a. Baseline and periodic follow-on medical physical examinations
   b. Regular physical training
   c. Annual physical fitness evaluations
   d. Education (training)

2. Reactive Measures
   a. Early recognition/treatment of ergonomically-based disorders
   b. Access to qualified medical care
   c. Alternative work
   d. Rehabilitation/"Work hardening"
   e. Workers' compensation

Written procedures should be established for both parts to form a medical management system. These written procedures should be reviewed and updated at least annually. Updating is necessary to keep abreast of rapid developments and changes in the field of occupational medicine.

Medical Management System

Supervision. Ideally, the medical program should be supervised by a physician, or an occupational health nurse with training in ergonomics and hopefully one who is familiar with the rigors and specific physical requirements of both fire fighting and emergency medical operations. If this is not possible, particularly for smaller or volunteer departments, the department should establish a relationship with an occupational medicine clinic where similar services can be obtained.

Section 8-6 of NFPA 1500 requires that departments have a physician for guiding, directing and advising fire fighters and EMTs with regard to their health, fitness, and suitability for specific jobs. NFPA 1500 also requires that the physician:

- provide guidance on the management of the department's occupational safety and health program,

- be a licensed medical doctor or osteopathic physician qualified in occupations safety and health as related to emergency services;

- be readily available for consultation; and

- provide services on an urgent basis.

The appendix of NFPA 1500 states:

> *"Depending on the size and the needs of a fire department, the fire department physician might or might be required on a full-time basis. A fire department should have a primary relationship with at least one officially designated physician. This physician can serve as the primary medical contact and, in turn, deal with a number of other physicians and specialists. A large fire department can designate more than one fire department physician or might determine that a relationship with a group practice or multiple provider system is more appropriate to its needs. In any case, the ability to consult with a physician who is particularly aware of the medical needs of fire department members and who is available on an immediate basis should exist."*

Primary Issues. The following issues should be included as part of any ergonomic-related medical management system:

- injury and illness record-keeping (see **Chapter 8**);
- early recognition/reporting of ergonomically-based symptoms (see **Chapter 2**);
- systematic evaluation and referral;

- conservative return to work;
- systematic monitoring, including periodic physical fitness evaluations;
- personnel training and education (see **Chapter 6**); and
- access to health care providers.

One component of an ergonomics program that is often overlooked is the psychosocial well-being of employees. By integrating mental health services into the overall occupational health program, on-the-job stress, both ergonomic and psychological, can be reduced.

The sooner symptoms are identified and reported, the better the chances are for effective and inexpensive treatment. Proven methods for accomplishing this goal are beginning to emerge. They often include training sessions in which personnel are instructed to report problems. Another useful step is providing periodic physical examinations to personnel whose jobs involve CTD risk factors (which includes most line fire fighters and EMTs). Steps also should be taken to ensure that when personnel report symptoms, standardized diagnostic tests are given and standardized procedures are followed for treatment and referrals.

The primary goal is to treat any health problem at an early stage to avoid surgery or other more complex treatment. Furthermore, mechanisms should be established to ensure that personnel return to work only when ready, are assigned to jobs that are compatible with any restrictions, and are evaluated periodically to see that problems are not recurring.

Specific guidelines for preventative and reactive measures of the medical management system are described below.

Baseline and Follow-On Medical Physical Examinations

All personnel should have an initial health status interview and physical examination to provide a basis for comparison, if symptoms emerge later. Specific requirements for medical examinations are established in NFPA 1500:

- Prior to becoming members, candidate fire fighters are to be medically evaluated and certified by the fire department physician.

- Candidate and members who engage in fire suppression must meet the medical requirements in NFPA 1582, *Standard on Medical Requirements for Fire Fighters,* prior to being certified for duty.

- Members who engage in fire suppression shall have at least an annual medical evaluation and shall meet the requirements as specified in NFPA 1582.

NFPA 1582, *Standard on Medical Requirements for Fire Fighters,* dictates the frequency and make-up of medical examinations for fire fighters and other emergency response personnel. According to the standard, the *annual medical evaluation* consists of :

- an interval medical history,

- an interval occupational history, including significant exposures to different hazards,

- height and weight, and

- blood pressure.

NFPA 1582 requires that **annual medical evaluation** include a **medical examination** based on the age of the fire fighter or EMT, according to the following schedule:

- Ages 29 and under - every three years.

- Ages 30 to 39 - every two years.

- Ages 40 and above - every year.

Specific body areas and functions required in the **medical examination** by NFPA 1582 include:

- Vital signs;
- Dermatological system;
- Ears, eyes, nose, mouth, and throat;
- Cardiovascular system;
- Respiratory system;
- Gastrointestinal system;
- Genitourinary system;
- Endocrine and metabolic systems;
- Musculoskeletal system;
- Neurological system;
- Audiometry;
- Visual acuity and peripheral vision testing;
- Pulmonary function testing;
- Laboratory testing, if indicated;
- Diagnostic imaging, if indicated; and
- Elecrocardiography, if indicated.

Figure 7-1 shows a detailed medical evaluation/examination used by *some* departments.

NFPA 1582 distinguishes between Category A and Category B medical conditions:

- A Category A Medical Condition is one that would preclude a person from performing as a fire fighter in a training or emergency operational environmental by presenting a significant risk to the safety and health of the person or others.

- Category B Medical Condition is one that, based on its severity or degree, may preclude a person from performing as a fire fighter in a training or emergency operational environmental by presenting a significant risk to the safety and health of the person or others.

7-4

Figure 7-l. Example Tests During Medical Examination

1. Complete Medical History
   - Medical/surgical history
   - Family history
   - Allergy history
   - Review of body systems
   - Prior work/exposure history
   - Prior toxic involvement

2. Ophthamologic Screening
   - Visual acuity - near/far point
   - Depth perception
   - Color vision

3. Audiometry
   - Hearing thresholds for 500 to 8,000 hertz

4. Urinalysis
   - Specific gravity
   - Albumin
   - Sugar
   - Blood
   - PH
   - Microscopic examination

5. Vital Signs
   - Temperature
   - Height and weight
   - Blood pressure
   - Pulse rate

6. Electrocardiogram
   - Twelve-lead resting trace
   - Treadmill exercises cardiac stress test (persons under 35 yrs old - every 3 years, persons 36-44 - every two years, persons 45 or old - every year)

7. Radiology
   - Chest X-Ray, PA 14 x 17 (administered on pre-employment, every year for smokers and every three years for non-smokers)

8. Pulmonary Function Screening Test
   - Vital capacity
   - l-set forced expiratory volume

9. Hematology Profile
   - Hemoglobin
   - Hematocrit
   - Red blood count
   - White blood count
   - WBC differential count

10. Blood Chemistry Profile
    - Calcium
    - Phosphorous
    - Glucose
    - Urea nitrogen
    - Uric acid
    - Cholesterol
    - HDL cholesterol
    - Total protein
    - Albumin
    - Total bilirubin
    - LDH
    - SGOT
    - SGPT
    - GGT
    - Cholesterol/HDL ratio
    - Globulin
    - Triglycerides
    - Alkaline phosphatase
    - Sodium
    - Potassium
    - Chloride

Some departments also employ other measurements for member physical fitness. One department uses the following evaluations for assessing fitness in addition to health:

- Determination of body fat (using skin-fold thickness measurements or other methods);
- Revaluation of flexibility (using a sit and reach test - 18 inches);
- Evaluation of aerobic power through measurement of maximal oxygen uptake using treadmill; and
- evaluation of strength using determinations of muscular endurance (push-ups, sit ups, lat pulls) muscular power (vertical reach, vertical jump, and grip strength), and muscular strength (bench pressure, leg press).

Other departments believe that strength and aerobic capacity tests can be better predictors of the potential for low back injury if they simulate the job in question.

It is important that results from successive medical examinations be compared with prior examination results to determine potential health problems. The combination of person complaints and results from fitness testing are extremely useful in diagnosing potential ergonomic disorders.

Regular Physical Training

It has been known that muscles can be conditioned and strengthened, but nerves and tendons cannot. *If the job is causing problems in certain muscles groups, then conditioning by exercise or other means is a worthwhile idea.* If the job is causing problems in the nerves and tendons (such as in carpal tunnel syndrome), conditioning is not likely to help. Since the nerves and tendons in the body cannot be strengthened, other control measures must be used such as redesigning the task or moving the person to a different responsibility for reducing the strain on nerves and tendons.

There is, however, strong evidence linking exercise that strengthens back and abdominal exercises for improving overall fitness, to decreases in the incidence and duration of low back pain. The role of exercise is to strengthen muscles and increase flexibility and endurance. Nevertheless, to be successful, fitness programs should be somewhat structured and supervised. A department fitness program should address:

- fitness program and operating rules,
- management and staffing,
- available space,
- time allotment, and
- incentives.

Fitness Program and Operating Rules. A comprehensive fitness or physical training program is one which address all aspects of fitness. Often fitness programs are one part of an overall "wellness" program which is now be encourage in the fire and emergency medical services. *To be successful, the physical training must involve and obtain the support of all parties involved - the municipal or regional government, the department administration, labor,* *and all personnel.* The following rules should be applied to the development of a physical training program:

1. Physical training should be non-contact activities.

2. Standard physical training uniforms should be worn during fitness activities (as opposed to work uniforms) since use of this clothing not only identifies fire fighters and emergency medical personnel as engaged in physical training but is more comfortable and conductive to physical training.

3. Conduct of behavior during physical training should project a high level of professionalism.

4. Physical training should be pursued in a safe manner.

Some departments allow sports in place of physical training, such as basketball. It comes as no surprise that there are a number of departments with injuries due to such sports either through improper playing surfaces or aggressive play. In any case, physical training should be focused on developing specific muscle groups and increasing aerobic capacity.

Management and Staffing. There needs to be one individual (the fitness coordinator), who will have the responsibility of managing and focusing all actions to ensure that all goals are met. This position should be filled by at least a first line supervisor. A strong background in the health and fitness field, backed up by effective managerial skills will also be necessary. Other staffing may be needed depending on the size of the department. If at all possible, attempt to obtain a temporary, full-time position in meeting with local government when the overall ergonomic program is being established. The reduction in sick leave and long term medical costs can be used a justification for obtaining this staff. Other potential sources are to use light-duty personnel, utilizing persons on an overtime basis, or looking to local colleges and universities as a resource for knowledgeable, energetic, and inexpensive assistance.

Space. Questions arise as to where fitness evaluations will be held and where regular physical fitness training equipment will be housed. The answer to these questions will obviously be based on the resources available and the ingenuity of the fitness coordinator. Equipment should be located where it will be easily accessible, preferably at individual stations, Some stations may require remodeling, others have empty apparatus bays that can serve as temporary gyms, while others may have no room at all. For this reason, it is important to start early in the planning and development of stations to allot space for appropriate exercise equipment, Other alternatives are to establish strategically located fitness centers or to look at available local recreational facilities. For example, a department may be able to make arrangements with local schools or commercial fitness establishments.

Time Allotment. The fitness program should be an on-duty activity for career fire fighters and EMTs. A minimum of 1½ hours should be allotted per shift. Personnel who work a 40 hour work week, should be allotted 4½ per week. The fitness program can be performed on either an in-service basis or out-of-service basis. The out-of-service basis is the most desirable option, but not at the expense of sacrificing basic responsibilities to the community.

Volunteer departments do not always have this luxury and consequently may rely on participating individuals making their own efforts to obtain and perform regular fitness training.

Incentives. Recognition of a job well done can take several forms, but each incentive has the same goals. The goals are to encourage participation in the fitness program, reward outstanding achievements in fitness, and ultimately encourage fire fighters and EMTs to live a healthier lifestyle. The least expensive incentive program is recognition. Examples of incentives include:

- Publishing remarks about an achievement made by an individual, company, division, or shift (e.g., having a monthly feature in the department's newsletter recognizing positive achievements in fitness and health).

- Posting individual achievements on fitness evaluation components, such as treadmill times can be successful. They not only award the individual who performed the feat, but it establishes a goal for the next individual to achieve.

Incentives must be made to accommodate all levels of achievement. Personal best fitness achievements should receive special attention. Personal bests provide realistic goals for the top athlete to the weekend warrior, by encouraging personal achievement. The types of achievement awards can include certificates to T-shirts. How and what awards provided will depend on the budget that must be worked within.

Annual Physical Fitness Evaluations

Annual physical fitness evaluations can be useful as part of a comprehensive medical management system in the fire and emergency medical services. The physical fitness evaluations should consist of two segments: clinical and job-specific tasks. The clinical segment uses more traditional methods of testing fitness such as push-ups, pull-up, and sit-ups. The data collected can be used as an objective method to determine the success of the fitness programs. The information can also be used as a preventative medicine tool by identifying early on the potential areas of concern and by helping the fire fighter or EMT implement an effective, healthy intervention strategy into their lifestyle. The job-specific fitness evaluation also referred to as "minimum company standards" uses simulated fire ground operations as a basis for determining physical readiness of the fire fighter or EMT. **Appendix D** provides an example of a complete fitness training evaluation used by the Phoenix Fire Department which may serve as a model for developing a fire/EMS department evaluation protocol.

Departments should avoid fitness evaluations which are designed more for competitive value then they are for physical conditioning and training. While there is utility for carefully managed physical performance assessments which consist of several tasks intended to simulate fire fighting operations, these assessments should not be used to set criteria for continued employment.

Ideally, physical fitness results should remain confidential to the individual and the physical trainer and should not be used by the department administration for qualifying field performance. In this manner, the fire fighter or EMT has positive incentives to improve and can use results as a tool in identifying weak areas of his or her fitness. If results are not kept confidential, then negative incentives are produced by the fitness evaluation.

Health Education

In order for fire fighters and EMTs to make informed decisions on health issues, they must be educated on proper health. They should be informed on the health hazards which have historically plagued the fire and emergency medical services and be convinced that they are at risk. They also must understand that if they choose to, they can make changes in their work habits and lifestyles that will significantly reduce their risk of injury or long-term disorders.

Recruit training offers one of the best opportunities to educate and motivate fire fighters and EMTs to a healthier lifestyle. The fitness program in the context of a fire academy should take several different forms. In addition to baseline fitness testing, fitness training should be emphasized throughout the program. When there are a number of recruits, the group should be separated into two categories based on the results of fitness testing. Those with more than adequate strength and marginal aerobic power should be taught different exercises which emphasize greater endurance. It is well known that tired workers are more likely to make mistakes which results in injury, frequently sprains and strains. Those trainees with less than adequate strength and adequate levels of aerobic power should undergo training to increase muscular strength. Recruit training is the ideal time to educate fire fighters and EMTs on the benefit of maintaining adequate strength and aerobic power to help minimize injuries.

Early Recognition of Ergonomically-based Disorders

A fundamental part of any ergonomics program is to recognize the symptoms of possible CTDs and other ergonomically-based disorders to begin treatment before the problem worsens and potentially causes more harm or lost time. **Chapter 2** extensively describes CTDs, muscular-skeletal disorders, and other ergonomically-based disorders. Though this part of the program overlaps with training **(Chapter 6),** it cannot be emphasized enough for departments to encourage worker reporting of pain and other symptoms early, so that specific treatments can be offered to prevent development of potentially severe injuries.

Access to Medical Care

The medical management system is highly dependent on supervision by a qualified physician or occupational nurse. This central authority is also a starting point for referral to specialized medical care when needed. In essence, it is possible to establish a network of medical specialists in the community which are proficient in providing certain types of treatment and have the extra benefit of understanding fire fighting and emergency medical operations. This ensures that appropriate decisions are made in providing treatment which address the particular needs of the department.

Unfortunately, not all departments can afford an associated physician or occupational nurse. These professionals may have to be identified in the community and then a relationship established to centralize an understanding of department physical requirements. However, this may not be possible for volunteer departments which consists of individuals all having different forms of health care coverage. In the latter case, the individual needs to assume some responsibility for informing their health care provider of special needs associated with their occupation. In any event, it is important that a physician or nurse have some occupational medicine experience to be knowledge in prescribed treatment protocols and work recommendations.

Alternate Duty

One of the most helpful treatments for employees with CTDs is a transfer to alternate-duty or restricted-duty jobs. Affected personnel should be able to work and feel that they are making a valuable contribution to the department. Often, line personnel experiencing the onset of CTDs or other ergonomically-based disorders may be transferred to administrative functions during a period of recovery. Supervisors should encourage injured personnel that they are valuable members of the work force.

Alternate-duty or restricted-duty jobs can be divided into two categories:

1. Jobs for individuals upon reporting symptoms of CTD injury on the job, and
2. Jobs for individuals returning to work after an injury has occurred.

The purpose of the first type of alternate-type job is to provide relief for physically-stressed individuals to prevent the condition from becoming worse. The alternate-duty job should allow the individuals to use different muscle groups than the original jobs required. It is also important that the alternative jobs be meaningful: avoid alternate jobs that have an implication of punishment.

Once affected personnel have been assigned to alternative jobs, it is imperative that they are monitored frequently. The purpose of monitoring is to determine if the condition is deteriorating or becoming better. This might be the case if the alternate job involves using the same muscle group as the job that caused the injury. If the fire fighter's or EMT's injuries do not improve, then another type of treatment must be undertaken, or the injured person must be transferred to a different type of job.

The purpose of the second type of alternate-duty job is to ease injured personnel back into their regular jobs. The physician treating the injured persons defines the restrictions and the department finds positions that the person can still perform within those restrictions. If the injury and restrictions persist, the injured person probably cannot return to the original jobs. Here, job retraining in other skills may be necessary.

Another component of an ergonomics program is the management of restricted workers. Communication between the treating physicians and the workplace task force is vital, as they both may be able to play a role in specifying restrictions. Restricted workers should be placed on light duty jobs that do not include repetitive motions or abnormal positions. A list of light duty jobs in each department should be developed by the task force and department supervisors. This list should be shared with the treating physician. In addition, periodic review of work restrictions is necessary to manage workers properly. Educating all personnel involved with restricted workers will help ensure success.

Rehabilitation

Work hardening is a concept applied to the rehabilitation of injured personnel. Work hardening is usually done in a clinic as part of an injured individual's treatment, and serves two important purposes:

1. It helps to heal the individual's physical problem, and it helps to heal some types of psychological injuries. At the rehabilitation clinic, injured individuals are gradually given exercise and job-related movements that will help them return to work more promptly than if no treatment were provided. This course of treatment helps to heal the physical problems. In a work hardening program, injured individuals report to the clinic in the morning, as if they were reporting to their jobs. They receive break periods, a lunch period, and, in between, they have a disciplined schedule to follow, similar to the workplace. Injured personnel are praised for even small amounts of progress.

2. It also prevents personnel from staying at home and promotes their psychological well-being. Statistics show that the longer injured personnel stay at home and off the job, the less likely they return to work. A work hardening program may prevent an injured individual from becoming one of these statistics.

Workers' Compensation

As a consequence of some ergonomically-based injuries, time on-the-job is lost time during period of recovery. If the disorder is of an extended duration, further anxiety can be produced when financial considerations come into play. With most departments, this lost time is covered by some form of workers' compensation program. A primary goal of workers' compensation is to "provide sure, prompt, and reasonable in some and medical benefits to work-accident victims or income benefit to their dependents, regardless of fault." However, benefits might not equal prior income, especially for volunteer fire fighters. All states except for New Jersey, South Carolina, and Texas have compulsory rather than elective compensation laws, but volunteers are not necessarily covered by these. Volunteer coverage can vary greatly between and within states, and benefits for a volunteer might not be based on the volunteer's "regular" job. **Appendix E** provide a list of general policies by state.

Instruction and training in an ergonomics program must include information about the financial consequences of long term ergonomically-base injuries. All fire fighters and EMTs are affected by the resulting high costs of medical coverage, and the injured fire fighter or EMT is affected even more directly by his or her (likely) reduced income.

Other goals of workers' compensation that parallel those of an ergonomics program are to "encourage maximum employer interest in safety and rehabilitation through appropriate experience-rating mechanisms," and to "promote frank study of causes of accidents (rather than concealment of fault) - reducing preventable accidents and human suffering." Information from workers' compensation claims is another resource for the ergonomics team for use in identifying and quantifying ergonomics concerns.

CHAPTER 8 - REPORTING PROCEDURES

According to the Occupational Safety and Health Administration (OSHA), ergonomic disorders are defined as "disorders of the musculoskeletal and nervous systems occurring in either the upper or lower extremities, including backs. These may be caused or aggravated by repetitive motions, forceful exertions, vibration, sustained or awkward positioning or mechanical compression of the hand, wrist, arm, back, neck, shoulder, and leg over extended periods or from other ergonomic stressors".

Although job-related injuries are referred to in a number of ways in the current literature (e.g. cumulative trauma disorders, repetitive strain injuries/disorders), OSHA clearly prefers the use of musculoskeletal disorders (MSD) as the descriptive term covering all occupational injuries/disorders. Current injury record-keeping in the fire service tends to concentrate on the acute or instantaneous variety with which fire fighters and Emergency Medical Services (EMS) personnel are quite familiar. Thus, *one major change that will be required of the fire and emergency medical services as ergonomic programs are started will be the inclusion of a new category: ergonomic injuries, plus documentation of the jobs or activities involved.* Manufacturing industries have for some time been required to keep records for all MSD's via a standard form (OSHA-200 form).

If job-related injuries go unreported until the ability of the affected individual to perform a given task is impaired, the results are time lost from the job, extensive medical or physical therapy treatment, and long recovery periods. The high costs associated with medical insurance and/or worker's compensation claims offer additional evidence for the advantages of ergonomic intervention.

Strategies designed to prevent or reduce ergonomic injuries on the job require sensitive and verifiable methods for identifying, reporting, tracking, and analyzing the data collected. Such systems must provide ways to target tasks that result in a disproportionately high number of reportable injuries. *A reportable injury may, for example, be defined as: anything that requires treatment beyond first aid.* The system must document injury type and location and enable users to determine whether the injury was ergonomically-related, or not. Furthermore, the system should track the disposition of the case for a documented period of time following the incident. An injury reporting system must be tailored to the specific needs of a department, but it may also contain the necessary elements to accomplish ergonomic goals. High among these goals is to identify jobs that produce ergonomic injuries so that early intervention can be initiated. Jobs that infrequently produce injuries may also be ergonomically unsound. These may well be modifiable, but will have lower priority.

The only way the level of risk associated with a given task can be quantitatively assessed over time is through accurate record-keeping. Logically, this requires that injuries must first be reported. Determining when to report and what to report are critical to the development of a useful database. Historically, some industrial workers (and fire fighters) have been disinclined to report minor injuries or (chronic) pain because of fear of supervisory reprisals, peer disapproval, or layoff. Often, injury was believed to be the individual's fault.

Of course, in some cases, people do foolish things, resulting in injury to themselves or others. Reporting systems must therefore be able to distinguish such incidents from those that may be ergonomically-derived. Ergonomics cannot correct human frailty, but good ergonomic design can help persons to act in ways less likely to produce a bad result. Fire fighters and EMS personnel often must operate under high stress conditions, or in hazardous environments upon which ergonomics can have only an indirect effect, but high skill and awareness levels can minimize unnecessary accidents. Early intervention into injury producing jobs is reactive ergonomics. With practice, a department will learn to react quickly and proactively.

The initial step in any proactive reporting system in traditional industry is the training of employees in the recognition of the signs and symptoms of an impending job-related disorder. Such training and communication at all levels of personnel is mandatory in order for any ergonomic program to be successful (Suggestions for these program elements are further discussed in **Chapter 6**).

The box below summarizes the basic components of a reporting system which captures the information and allows its effective use as part of an ergonomic program. It is not enough to simply collect and record information. The analysis and subsequent use of reporting provides the collected data is the activity which provides departments the ability to assess ergonomics hazards and to determine the effectiveness of proposed solutions.

BASIC COMPONENTS FOR ESTABLISHING A SUCCESSFUL REPORTING SYSTEM

- *DEFINE* what constitutes a reportable occupational injury and/or disorder.

- *IDENTIFY* lines of responsibility, i.e., to whom is an injury reported and what is he/she to do with the information?

- *TRAIN* personnel to heighten awareness: Recognize the importance of reporting an injury/disorder. Learn to identify conditions that may be occupationally related.

- *REASSURE* individuals that reporting an injury/disorder will not be used to penalize individuals.

- *RECORD-KEEPING:* Establish data needs, then select software that will supply the desired information. Train staff in the use of such software.

- *FOLLOW-UP:* When a hazardous job/activity is identified, establish means to track the action taken to correct the problem.

Current Reporting Practices

A review of fire and emergency medical services practices in 1995 revealed that no surveyed major departments around the country had an ergonomic program in place. Nevertheless, many departments apparently use some form of injury reporting and record-keeping at the present time. The primary reason and major factor affecting the designs of current injury reporting/record-keeping systems is related to insurance and workers' compensation. These reporting formats and methods for recording information are typically not ideal for ergonomic purposes. Also, many aspects of the content or even the number of forms used may depend upon local or state requirements or regulations:

- For example, in one department, the forms are not specific to the fire or emergency services, but are either required by a county-wide risk management division or a statewide worker's compensation commission,

- In another case, a workers' compensation form plus a statewide Department of Fire Programs, Fire Incident Reporting System form are used. The latter form apparently is restricted to injuries incurred in transit or at the fire and/or medical emergency scene. In addition to general background and personal data and descriptions of the activity, action taken, and medical data, the form records considerable information relative to clothing and personal protective equipment use. While much of the information recorded is very useful, the form is clearly not directed at resolving ergonomic issues unless they involve the turnouts, boots, helmets, gloves, or breathing apparatus.

- Still other departments have very brief injury reporting systems which typically provide only enough data to enable collection under workers' compensation.

On the positive side, this form which was clearly derived from the National Fire Incident Reporting System (NFIRS) Handbook, is oriented toward easy databasing.

ABOUT NFIRS

NFIRS is the National Fire Incident Reporting System developed by the U.S. Fire Administration for providing a database of fire incidents and casualties. As such, the NFIRS database provides a means for analysis of fire incident data for identifying patterns, problems, and trends. Such analysis can be used by fire departments to help define current problems, predict future problems, and set appropriate budgetary priorities. NFIRS information is also used at the state and local levels in the evaluation of issues such as fire fighter injury and workload. State participation in NFIRS is voluntary.

Table 8-1 provides an outline of the kinds of information contained on most common fire and emergency medical department injury reporting forms. The content and format of forms are quite different. Some rely almost entirely on hand written entries, while others contain blanks with coded entry suitable for databasing. As noted, the forms reviewed were not specifically designed for extracting ergonomic information. As evidenced by the outline above, the information concentrates on accidental, instantaneous, or traumatic types of injuries, rather than on job or occupationally related musculoskeletal disorders. Some minor exceptions were found on the recently developed form used by a northwestern fire department. For example, "carpal tunnel", "overuse syndrome" and "tendinitis" are listed under the *Nature of Injury* section, while another entire section is entitled *Cause of Injury or Occupational Disease*. Under the section *Contributing* Factors, several entries are typical of ergonomic concerns, *e.g.,* body mechanics, fitness for work, physical condition.

Recommendations for an Ergonomic Reporting System

In the developing an ergonomically-based injury recording form for the department, the example forms provided in **Appendix F** should first be reviewed. Then a decision can be made for which items from the outline in Table 8-1 or the sample forms in the appendix can be chosen. *Recent examples of accidents involving injuries should be considered to determine if all the relevant information has been captured which allows an understanding of the nature, cause, and corrective actions for a particular incident.* If other forms which are required by the city, county, state, or insurance company are currently computerized, do not rely on sharing information in order to complete your ergonomic database. The injury reporting data should be complete unto itself.

In addition to those data items selected from the summary list provided in Table 8-1, ergonomically-oriented injury records should include the following data in some form:

1. Add the following fields of information to the "INJURY: DESCRIPTIVE HISTORY, *cause of injury*" subsection (D3): repetitive motion, awkward posture, forceful exertion, prolonged exertion, frequent/heavy lifting, contact stress, local/whole body vibration, environmental conditions (hot/humid, cold/freezing, poor lighting, toxic agent), faulty equipment (use problem, design problem, modification made, alternate supplier), other

2. Subdivide the "INJURY: MEDICAL HISTORY, *type of injury*" (E2) subsection into:
   a. accidental/traumatic/acute-burn, abrasion/contusion, fracture, laceration, smoke inhalation, sprain/strain, tendinitis
   b. occupational/job related-cumulative trauma disorder, repetitive stress injury, musculoskeletal disorder

Table 8-1. Typical Entries for Department Injury Reporting Form

A. *ID* - department, case number, distribution

B. *PERSONAL* - individual's name, sex, age

C. *ADMINISTRATIVE* - rank, unit, shift, time in service, previous injuries, other forms filed (worker's compensation, insurance, NFIRS)

D. *INJURY: DESCRIPTIVE HISTORY*
  1. location - at station, fire scene, aid scene, accident scene, in transit, false alarm.
  2. activity (at time of injury) - operating, overhauling, taking-up, training, exercise.
  3. cause of injury - vehicle accident, burn, fall, fire/smoke, explosion, lift, push, pull, bend, slip, needle puncture.
  4. severity - precautionary, treated-no time lost, treated-time lost, hospitalized, fatal.

E. *INJURY: MEDICAL HISTORY*
  1. body part injured - head, face, arms, hands, trunk, legs, feet
  2. type of injury - burn, abrasion/contusion, fracture, laceration, smoke inhalation, sprain/strain, tendinitis
  3. first treatment - at scene, in transit, at station. Who performed the treatment?
  4. level of action - first aid only, Emergency Medical Technician (EMT) treated, doctor treated and released, doctor treated and sent home, doctor treated and hospitalized.
  5. estimated recovery period
  6. rehabilitation requirements

F. *CONTRIBUTING FACTORS*
  1. hour on duty
  2. time of day
  3. environmental conditions
  4. equipment failure/problem (see below)
  5. practices
  6. procedures

G. *CLOTHING/EQUIPMENT USED AND DESCRIPTION OF PROBLEM*
  1. helmet
  2. gloves
  3. turnout
  4. boots/shoes
  5. breathing system
  6. other

H. *EXPLANATORY INFORMATION*
  1. written description of accident
  2. written description of injury
  3. written description of treatment given
  4. time elapsed between injury and treatment

3.   Add the following new sections:

ACTION/CONTROLS
1.   risk/hazard checklist
            checklist performed?
      b.   OSHA score per proposed draft ergonomic standard?
      C.   injury unavoidable?
2.   engineering controls - physical change to workstation, equipment, or materials
3.   administrative controls - redesign work methods/requirements in the future

MEDICAL RESTRICTIONS (post injury)
1.   alternate duty assignment (estimated time)
2.   light duty (estimated time)
3.   manual material handling limits (estimated time)
4.   other (specify)

ERGONOMIC RELATIONSHIP
1.   Was this injury ergonomic?
2.   Is cause correctable?
3.   If correctable, what action was taken?

FOLLOW-UP INFORMATION
1.   <u>no</u> detectable effect on outcome of injury
2.   <u>mild, moderate,</u> or <u>strong</u> effect on outcome of injury

REPORTING CUMULATIVE SUMMARY
1.   no injuries of this type per (month/six months/year)
2.   total injuries (per month/six months/year)

Department records should be integrated but not combined with required state/local data forms.

Results of ergonomic actions should be published and this information publicized to personnel and management. This information can be passed down to everyone in the department by bulletin board postings, weekly station safety meetings, or in department newsletters.

Developing a Database for Analysis of Injury Data

The major purpose of developing an injury database is to enable correlation of high risk jobs with the injury rate:

●   In small departments, a database may simply be developed from handwritten injury recording forms.

●   In larger departments, a computerized system will probably be necessary to identify true problem areas.

In either case, once identified, high risk jobs become the priority candidates for change or at least close surveillance by the ergonomic or safety committee team members selected for this purpose. Upon determination of the correlation between high and low risk jobs, and high and low injury rates, the database must be so constructed as to permit further analysis. Such analyses will require database software that permits examination of multiple variables simultaneously. In addition, the capability of the software to provide summary graphic presentations of the data will be very helpful in defining difficult or questionable problem areas.

One of the problems in developing an ergonomic program has been the difficulty of ascertaining from current injury records whether departments are actually experiencing ergonomically-correctable injuries. Probably most departments, regardless of size, have such injuries, but the reporting system does not allow the extraction of pertinent data. Specifically, it is likely that the ergonomic injuries are hidden (unintentionally) by being spread out through several other categories. For example, the combination of *Injury Type: sprains/strains* with *Injury Location: lower back* probably indicates an ergonomic injury in many cases. But there is no way to tell who the person was or what he/she was doing at the time, even in the most sophisticated reporting system generally observed throughout the fire and emergency medical services. The problem appears to be that the software is programmed to provide data analysis output in fixed ways, but will not permit the user to access variables so that multivariate interrelationships including statistical analyses can be examined (such as determining how many people suffered lower back injuries with time lost while removing ladders from fire apparatus). In other cases, different types of variables are used, or variables are not encoded in ways that make analysis convenient *(e.g.,* use written information when only digital information can be summarized.)

Thanks to the nominal cost of powerful portable computers and software packages, it is easier to maintain ergonomic program components and injury data in a database for those departments which can use computer resources for this purpose. Most departments, regardless of size, currently use computers for some kinds of record-keeping. The format used is not critical nor is it likely that all departments would find a standard format suitable for their specific needs. It is important, however, that certain minimum information be collected, and that the format adopted is compatible with any software chosen. The first thing to be considered is the need to meet record-keeping requirements found in NFPA 1500 (Section 2-7).

Once the data have been entered in a database format and verified, a great variety of possible breakouts of the data can be run. For example, the following types of cumulative information should be available from the selected or developed software:

- number of injuries by month and year, by station, and by full department
- number of injuries by type, name, and severity
- number of injuries by type, location and activity at time of injury, by time of day, shift, age, sex and length of service
- severity index-injury type, days lost, number of members
- ergonomic index-number of injuries related to ergonomic causes, type of injury, days lost

- follow-up data-number of injuries, by total or by type that required rehabilitation
- number of injuries leading to change in job/activity, engineering and/or administrative
- incidence rate (ratio of the number of injuries to the frequency of task related to the injury is performed)

The above are but a few examples of the kinds of summary or cumulative information that should be available from your database. In addition, the ability to run basic statistics, frequency plots, and other graphics is very useful, particularly in making presentations to management for justifying safety or tactical changes.

Modern software packages permit the user to custom-design forms, design data fields, enter new data, edit old data, and perform various types of analyses of the data. Easy-to-use pull-down menus and/or Windows versions are available for most applicable software. The use of a spreadsheet format for the data fields facilitates individualized types of analyses via existing statistical options or graphic presentations. In some cases, simple batch file programs may achieve the desired results. In any case, in developing your ergonomic injury data form, be sure that personal information, as well as all injury data, are coded in a logical digital or alphabetic form which will simplify desired analysis of any or all variables. The database format should be constructed so that *univariate* and *multivariute* statistical analyses (see box below) can be performed.

TYPES OF INJURY DATA BASE ANALYSES

Univariate Analysis - An analysis of data which permits the determination of the number of injuries associated with one variable. For example, the total number of injuries involving lower back problems (type of injury).

Multvariate Analysis - An analysis of data which permits the determination of the number of injuries associated with two or more variables. For example, the total number of injuries involving lower back problems (type of injury) due to removing hose from apparatus (cause of injury) resulting in lost time (result of injury).

Small departments may have access to a portable computer (PC) with enough hard drive capacity to handle the injury record-keeping for up to a year at a time. The data could be transferred to tape or floppy disks for permanent storage each year for those departments handling large quantities of injury information. Probably, at least one fire fighter or EMT in each department will be an experienced PC user and eager to put his or her skills to work for the department. Such a person can provide the impetus for starting the process of record-keeping. He or she can pursue the task of finding the appropriate software, data entry, and analysis techniques. Alternatively, if a member cannot be found, a person from the community might voluntarily set up a system for smaller departments.

It may be that in small departments (less than 100 members), the injury rate is so low that the use of a computer may not be necessary. Documentation of reportable injuries over a year or more may be accomplished more efficiently and with less cost through the use of handwritten forms. Such forms, like those used for computer entry, could be designed around the examples found in **Appendix F** with the addition of the ergonomic information recommended in this report. The form, once designed to fit your specific needs, can be reproduced and printed commercially at a nominal cost. Whether or not to use a computer can be a choice made by the department.

Most intermediate and large departments currently maintain some form of computerized record-keeping of injuries or other data, so that moving to a more ergonomically-oriented record system may be a simple matter of modifying the existing entry format to include the new information. A wide variety of software packages that are currently on the market can be selected to handle the development of your database. Manufacturer-supplied software basically comes in two forms: complete systems sometimes referred to as *turn-key systems,* which may be installed and put into use without user setup, and *modular systems,* which consist of both hardware and special application programs. Turn-key systems require the user to do no programming. In some cases these systems are supplied with read-only memories (ROM) which cannot be accidentally destroyed and which have been specifically programmed to perform a fixed sequence of operations. Modular systems, on the other hand, can be expanded to special purpose systems tailored to individual needs.

The size of the department, will in some, cases determine the type of software system in which you should invest. Many software companies will supply some free consulting help. Special workshops developed by the software suppliers are frequently available in your area to help those having problems with some aspect of their program(s). Almost all communities will have available computer consultants or companies that will write software specific to your needs. First, the ergonomics committee or other responsible group must decide exactly what your needs are so that the cost can be estimated. One or two individuals appointed by, or volunteering from, the ergonomic committee may be sufficient to select the method of approach. Once your needs are identified, at least some amount of consulting by a database expert before your purchase is made may prove to be very beneficial. In large departments, management information system specialists may already be on the staff. If so, involve them with the ergonomic team early in the development phase of the program.

TEN STEPS IN THE DEVELOPMENT OF AN INJURY DATABASE

1. Establish the extent of the database: it should cover the whole department with options to break out unit data. The format of your system should be compatible with that of local, state, and federal forms.

2. Create a team (as part of the ergonomic committee, safety committee or other responsible group) that is responsible for overseeing the development of your database.

3. This team should do the research necessary to determine the information to be contained in the database. Develop a taxonomy and data coding strategy.

4. Establish goals, requirements, and uses of the database. Establish criteria for inclusion of each datum.

5. Select the injury measures to be used. All data must ultimately be translatable into a common measures.

6. Select a commercially available software package that fits the anticipated needs. The package selected should be user friendly and easily learned by novice users. It should be easily adaptable to changes. Alternatively, contract with a data base consultant to create a custom program. A capable volunteer may be found within the department, or in the community.

7. The system should include, or be compatible with, graphics, statistical programs, and database translation software.

8. Organize the program structure to be compatible with record forms.

9. Exercise the program with trial data entry and modify as necessary.

10 Establish lines of responsibility for maintenance of the database.

The successful establishment of any new program requires the formulation of a general plan of attack. Strategies will vary from department to department, but the series of steps outlined below constitutes a basic model which can be used as a point of departure.

- Undertake the preliminary ground work necessary to gain support' and understanding for the program from members of the department, local government officials, and members of the community.

- Schedule a kick-off meeting called and presided over by the chief to explain and launch the program.

- Establish an ergonomic committee (a rotating membership is preferred).

- Define committee's rules and roles.

- Develop immediate goals and create a long term plan.

- Establish communication pathways and lines of responsibility.

- Develop awareness of ergonomic principles among all levels of personnel.

- Identify tools and resources - guidebooks, manuals.

- Develop record-keeping techniques and forms.

- Integrate forms/software/analysis of data.

- Learn to recognize, analyze, and redesign high risk jobs.

- Train members in safe work practices.

- Develop proactive methods for modifying high risk activities through risk assessment, symptom surveys, and follow-up/record-keeping procedures.

- Implement program evaluation and "sales" methods.

Gaining Support

Experience in industry has confirmed that the involvement of persons at all levels is vital to the success of an ergonomics program. Because of the quasi-government nature of fire departments, total involvement from top to bottom is perhaps even more important than in industry. Thus, the first steps of the department's plan should be to prepare the way for an ergonomics program:

1. Pretrain one or two representatives from the department to serve as spokespersons in "selling" the concepts both internally and to the larger community. The department physician and/or the medically responsible person on staff may be the best choice along with an EMT.

2. Organize a series of meetings across the department so that all shifts have an opportunity to attend an introductory seminar. If funds permit, also use a local consultant or expert to present concepts. If funds for such purposes are limited, a local expert may be willing to contribute services. Alternatively, a nearby industry that has a successful ergonomic program will usually be pleased to share experiences. Have available visual materials to help explain ergonomics.

3. Schedule a kick-off meeting with all levels of personnel well represented. Invite council members, commissioners, the mayor, a safety officer from a local company with a successful program, and interested local citizens to attend. Present the idea of establishing an ergonomic program using simple, but convincing arguments: the reduction of injuries, and of person/days lost to injury; the reduction of worker's compensation costs and medical insurance costs; the need to be prepared for mandatory state or federal regulations which are likely in the future.

4. Conduct a cost-benefit analysis using the guidelines provided in **Chapter 10.**

Establishing an Ergonomic Committee or Responsible Group

A well-functioning ergonomic committee is obviously the linchpin of any successful program. Its job is to develop an ergonomic plan, see to it that the program is executed, and provide continuing guidance and coordination of its activities. Among its specific tasks will be to:

1. Determine existing problems.
   a. perform a job risk analysis
   b. review layouts of all buildings and facilities for hazardous features

2. Develop a reporting and record-keeping strategy.
   a. select needed data forms
   b. select needed software
   c. train personnel to recognize and report ergonomic risks and injuries

3. Set up a training and continuing education program for all members of the department. Identify and obtain needed tools and resources, e.g., guidebooks, manuals, videos, workshops, consultants.

4. Evaluate new tools, machinery, and processes for reducing ergonomic hazards.

5. Serve as liaison between all levels of personnel and between the department and relevant government entities.

6. Develop and maintain a tracking system for job analysis and the all-important activities associated with modifying jobs, equipment, or environmental factors that present avoidable risks.

Individuals selected for the ergonomic committee should have demonstrated interest and qualifications. At least one member should be medically trained (minimum of EMT or paramedic). If the department has a staff physician or a contracted physician, he/she should be part of, or chair, the team. Membership should also include a top administrative officer and representatives from the line fire fighters and EMT's. Depending upon department size, the committee may include four or more members. Union members, if they exist, must be included. Each committee member should be assigned an area of responsibility. Suggested responsibility for the committee are provided in Table 9-1.

Integration of Ergonomics in Existing Safety Programs

Safety training is routine in the fire service. The integration of ergonomics into an existing department occupational safety and health program should be a smoother process in the fire and emergency medical services than in any manufacturing industry. Specifically, personnel in both services, either career or volunteer, are usually highly motivated individuals, and maintain a good level of physical fitness in order to perform their jobs. Frequently, in small departments or departments with largely volunteer personnel, many individuals act in both fire fighter and EMT roles depending on need. Many of them may have served on an ergonomics team at their place of employment. Such prior training should be taken advantage of.

The normal training of fire and EMS personnel is not like unlike that needed for ergonomics. Probably the major differences are as follows:

- Most job-related injuries that an ergonomic approach intends to minimize, occur over time and are not of the acute or instantaneous type with which fire fighters and EMS personnel are familiar.

- Jobs will be evaluated on the basis of risk factors associated with the number and intensity of certain biomechanical actions on the job.

- Risk factors are usually categorized by the area of the body involved.

- Many acute injuries may have been caused in part by poor design or by misuse of a tool or piece of equipment. Ergonomics and safety training both address these problems.

- Many existing safety and fitness training programs currently teach lifting techniques in order to reduce the incidence of back injury. Ergonomics and safety training both address this classic and costly problem.

- Internal record keeping practices may be new or different.

Table 9-1. Recommended Ergonomic Committee Member Basic Responsibilities

A. **Health and safety representative**
1. evaluate ergonomic interventions for safety
2. design and supervise injury record-keeping
3. educate personnel
4. guide job intervention schedule
5. provide continuing education in cause and treatment of injuries
6. encourage workers to use implemented ergonomic improvements

B. **Line fire fighters and EMTs**
1. apply knowledge of jobs to the ergonomics solution process
2. ensure that recommended modifications are practical and stay on schedule
3. interface with co-workers to ensure their cooperation during job analysis
4. provide insight on job analyses
5. help establish broad understanding of ergonomics process among fellow employees

C. **Supervisory representative (Lieutenant, Captain, Chief)**
1. interface with all sections of the department
2. coordinate schedule information
3. act as resource for conducting job analysis
4. facilitate administrative actions

D. **Engineering/maintenance representative**
1. coordinate engineering problems l
2. provide input on future needs
3. provide engineering solutions
4. facilitate engineering solutions

Putting the Program into Practice

As discussed above, application of an ergonomics-based perspective must begin with the ergonomic committee, all of whose members should receive training in all areas. Later, depending upon department size, teams can become more specialized. Supervisory personnel, fire fighters, and EMTs on the team will learn which areas fit their skills best, e.g. reporting and record-keeping, training, and risk assessment. All personnel should receive basic biomechanical/medical awareness skills training.

One important objective of the program at all levels is to instill an awareness of the possibility of problems even though injury records do not reflect high risks. Often a complex analysis may not be needed, since a history of injury on the job will make obvious the need for change. In other cases, simple changes in a particular tool or item of equipment would result in an ergonomic benefit. Other jobs or activities may be too complex for instantaneous analysis. These should be subjected to symptom surveys or risk analyses, as part of the gradual review of all identified jobs.

Recognizing that fire fighting and the EMS are considerably different from traditional industry and that existing ergonomic programs may be quite new, evaluation may be difficult. Existing evaluation criteria are typically based upon those found useful in much different industries. In industry and in general, the success or failure of a program is frequently assessed by management in terms of economics:

- increased net production (efficiency due to lower number of quality control rejects),

- reduced incidence of days lost from work,

- reduced medical insurance cost,

- reduced worker's compensation cost, and

- increased 'esprit de corps.'

As can be seen, all but the first of the items on the above list apply to the fire fighting and EMS professions. A successful program anywhere must show a measurable reduction in injuries, severity of injuries, and work time loss after the program is put into place. The problem may be that such a large proportion of fire fighter injuries are not ergonomically-related. For example, the major proportion of burns, slips and falls, smoke inhalation, cuts, abrasions, and contusions are suffered at the fire or accident site where the environment and the general circumstances are never the same and frequently outside the person's ability to control, Such injuries can only be classified as ergonomic if they are the result of faulty or poorly designed equipment, or their severe misuse. Such equipment can, of course, be modified. Ergonomics teaches what kind of modification is best from a human factors viewpoint. Fire fighters and EMS personnel are generally extremely well trained, physically fit, and strong individuals who typically know how to get the most out of their equipment. But under the stress of a major fire or accident, mistakes are made. In some cases an "ergonomic awareness" will help.

Recommended Evaluation Criteria

The fact that ergonomics is new to the fire service may require the development of brand new criteria by which to determine the long-term success or failure of a program. In the long term, the effectiveness of an ergonomic program can be measured at the local level regardless of the size of the department by the following criteria:

1. Health statistics
        reduced injury rate
    b. reduced injury severity
    c. reduced overhead costs

        d.   reduced medical/workers' compensation costs
        e.   reduced time loss

2.   Program statistics
        a.   increasing number of ergonomic hazards identified by line personnel, as opposed to trained persons
        b.   increasing frequency of ergonomic fixes approved by management

3.   General
             improved, safer work environment
        b.   better equipment developed/purchased
        c.   improved work practices

Finally, a successful ergonomics program will have a number of unquantifiable benefits, such as increased morale and improved response efficiency. These characteristics should be evident when a successful program is underway:

- better in-house *esprit de corps,*
- improved administration/line/labor personnel relations,
- improved department/community government relations,
- feeling of enhanced safety at scene,
- better communications between divisions, and
- better communications between specialties/specialists.

Cost/Benefit Analyses

Perhaps the best method for determining the effectiveness of the program is to conduct a cost-benefit analysis. ***To justify implementing or maintaining an ergonomics program, a department may need to quantify the costs and benefits of the program and compare them to the costs with no program in place.*** To support implementation of a program, estimates or cost projections may be used for various aspects of the program. For a program already in place, actual data from past years should be used.

Table 10-1 depicts a sample worksheet for departments attempting a cost/benefit analysis. Table 10-2 includes descriptions of categories included on the worksheet. Because the definition of the term "ergonomic-related" can be quite subjective, a rigorous analysis may not be feasible; determining ex post facto which injuries are actually ergonomic-related may not be possible, However, cost trends should be evident. Bar graphs may be a useful tool for comparing "before and after" figures for various cost categories.

First, the department will need to define what is considered an ergonomic injury. For convenience, selected types of injuries, such as sprains and strains, and specific body parts, such as the back and shoulder, may be targeted. While not all of the targeted injuries may be truly ergonomic-related, it is probable that a large number will be ergonomic-related. As long as definitions are consistently applied, reasonable cost estimates should be attainable for comparison purposes.

Table 10-1. Cost/Benefit Analysis: Data Summary Worksheet

Category	A. Pre-Program	B. Program Set-Up	C. During Program	Difference = (B+C) -A
Injuries				
Lost-Time Injuries				
Lost Days				
Restricted Duty Days				
Medical & Rehabilitation costs				
Worker's Compensation Settlements				
Employee Replacement cost				
OSHA Fines				
Consultant Fees				
Equipment				
Training Materials				
Labor- Ergonomics Team & Staff Training etc.				
TOTAL COST				

Table 10-2. Category Descriptions for Cost/Benefit Analysis Data Summaries Table

Category	Description/Special Notes
Injuries	Apply consistent definition of "Ergonomic Injury." May be also useful to computerize data for all injuries for comparative purposes
Medical and Rehabilitation costs	Include physician and therapist fees; hospital, clinic, lab fees; Rehabilitation costs, including equipment, facility use
Worker's Compensation Settlements	Include Worker's Compensation settlements for permanent and temporary disabilities; May need to adjust for fact that often these settlements are for injuries in prior years
Employee Replacement Cost	Include costs of hiring and training replacement
OSHA Fines	Consider the possibility of an OSHA citation under the "General Duty Clause" or future OSHA standard on ergonomics; While this cost is difficult to predict, $20,000 has been suggested as an "educated guess"
Consultant Fees	Include one-time fees from outside consulting firms for setting up programs as well as fees for managing on-going ergonomics, managed care, and/or fitness/rehabilitation programs
Equipment	Include exercise equipment and special equipment, tools, personal protective equipment purchase for ergonomics reasons; Include costs of retrofitting apparatus or modifying facilities
Training Materials	Includes costs of books, videos, equipment used for training
Labor - Ergonomics Team, Staff Training and Counseling	For team: Include time spent to research options, develop program, receive outside training, and manage program; For staff: Include ergonomics training, periodic evaluations, and counseling for individual ergonomics needs; Include exercise time if not already incorporated into shift

Basically, the cost/benefit analysis requires determining pre-program injury costs for a specified time period, and then comparing these to the costs of implementing an ergonomics program and the injury costs with the program in place for a similar time period.

Appendix G includes an evaluation of cost effectiveness by the Riverside (California) Fire Department.

It should be noted that a single catastrophic injury could skew injury data. In this case, to obtain a clearer picture of injury and costs trends, the analysis could be calculated twice, first with the catastrophic injury retained and then with the injury omitted. As an example, the information included in **Appendix G,** the Riverside evaluation of its physical fitness program, was entered into the sample cost/benefit analysis data summary worksheet. Data not included in that report is indicated by "NA" in the table. Statistics were not adjusted to omit major injuries. This worksheet illustrates a reduction in lost-time injuries of 24% (13 injuries) over a three-year period, although total injuries increased by 11% (18 injuries). Program savings for the three-year period were $194,282, or 32 %.

Costs Associated with Injuries

Costs associated with injuries may fall into six categories:

1.   Time (labor costs)
        a.   lost work time
        b.   training time for job reassignment

2.   Medical Costs
        a.   doctor, therapist, hospital, etc. costs
        b.   rehabilitation costs

3.   Personnel Costs
        a.   temporary replacement cost
        b.   employee replacement cost, including new training
        c.   cost of excess pay if reassigned to job usually performed by lower paid employee

4.   Workers' Compensation Distributions and Settlements

5.   Fines from OSHA Citations (industry only)

6.   Intangibles
        a.   loss of morale
        b.   negative publicity or negative public image

While it may not be feasible to quantify all of these costs, estimates can be made for many of them. Departmental injury records can be investigated to determine medical and other costs for a given period. Workers' compensation settlement data can be examined to identify possible ergonomic-related cases. The cost of OSHA citations may be applicable under the

"General Duty Clause" if the physical demands of the job exceed ergonomic criteria established by NIOSH or other sources. One source suggests $20,000 as an acceptable "best guess" for a citation, although this is considered relatively high for a fire or emergency medical services department. Due to this uncertainty, inclusion of a citation figure in an ergonomics feasibility study is controversial and it is recommended to leave this entry blank.

While injury information can vary greatly between departments because of the highly unpredictable nature of the work, examples of individual department dam are presented in Tables 10-3 through 10-8 as a guide for departments beginning the cost analysis process. Table 10-4 gives the average lost time for certain body part injuries incurred by a Minnesota fire department. Table 10-5 includes examples of medical costs for certain types of injuries for the same department. Table 10-6 details cost and lost time for injuries identified as ergonomic-related for a Virginia fire department. The total costs for outside medical payments, injury leave, and temporary disability are listed in Table 10-7 for an Oklahoma department. Average workers' compensation settlements for this same department are listed in Table 10-8.

Costs Associated with An Ergonomics Program

The costs of designing, implementing, and maintaining an ergonomics program can be associated with three areas:

1. Outside Fees
    a. consultant fees for job analysis, program design, personnel training, and program implementation
    b. fees to outside fitness experts and fitness facilities for implementing and maintaining programs

2. Materials
    a. exercise/fitness equipment
    b. training literature and equipment
    c. replacement or retrofitting of furniture, apparatus, turnout gear, and other equipment
    d. possible station redesign

3. Time
    a. time spent by those responsible for implementing program
    b. training time
    c. exercise time (if not already allotted as regular part of shift)
    d. time for periodic evaluations and individual counseling

Departments may also incur costs for setting up managed medical and rehabilitation programs. Depending on the department size and demographics, costs for these can vary widely.

Table 10-3. Cost/Benefit Analysis: Data Summary Table for Riverside, California

Category	A. Pre-Program (1984-1986)	B. Program Set-Up (1986)	C. During Program (1987-1989)	Difference = (B+C) - A
Injuries	163		181	18
Lost-Time Injuries	5 4		4 1	(13)
Lost Days	229		111	(118)
Restricted Duty Days	N A		N A	NA
Medical & Rehabilitation Costs	$474,233		$256,441	($217,792)
Worker's Compensation Settlements	N A		N A	NA
Employee Replacement Cost	$122,232		$59,623	($62,609)
OSHA Fines	N A		N A	NA
Consultant Fees		$16,720	$19,665	$36,385
Equipment		$49,734	-0-	$49,734
Training Materials		-0-	-0-	-0-
Labor Ergonomic Team & Staff Training, etc.		NA	N A	NA
TOTAL COST	$596,465	$66,454	$335,729	($194,282)

Table 10-4. Average Lost Time for Injuries with Lost Time,
for a Minnesota Fire Department

Body Part Injured	Average Lost Time in Days per Injury, by Year				
	1990	1991	1992	1993	1994
Back	8.5	12	9	9.5	5.1
Lower Extremities	7.9	10.1	12.1	11	4.3
Neck	59	0	0	0	4.5
Trunk	15.5	5	32.6	23.5	13.9
Upper Extremities	11.7	3.3	4.6		

Table 10-5. Average Medical Costs for a Minnesota Fire Department

Body Part Injured	Average Medical costs per Injury, by Year*			
	1991	$1992	1993	1994
Back	$2094	$2336	$1061	$1555
Lower Extremities	$905	$1434	$2030	$3672
Neck	$96	$83.5	$550	$244
Trunk	$203	$5306	$11,154	$1254
Upper Extremities	$177	$726	$2030	$5400

*includes injuries with and without lost time, based on 1994 dollars

Table 10-6. Medical Cost and Lost-time for Specific Ergonomic-Related Injuries for an East Coast Fire Department (1993-1995)

Injury	Location	Cost	Lost Time
Ankle Sprain	EMS	None	2 hours
Arm Strain	Fire Scene	$3,201	3 hours
Arm/Shoulder Sprain	Fire Scene	$16,330	40 shifts
Arm Strain	Administration	None	4 hours
Arm Strain	At station	None	None
Back Strain	At station	$156	72 hours
Back Strain	Administration	$471	None
Back Strain	At station	$5	24 hours
Back Strain	EMS	None	None
Back Strain	Training	None	None
Back Strain	Fire Scene	None	None
Back Strain	At station	$379	9 hours
Back Strain	Fire Scene	$1377	None
Back Strain	At Station	$61	None
Back Strain	EMS	None	None
Back Strain	In responding	$506	4 hours
Hernia	At station	$3,459	35 shifts
Knee Strain	Fire Scene	$193	None
Leg Strain	In responding	None	48 hours
Neck Strain	Fire Scene	None	None
Neck Strain	Training	$324	None
Neck Strain	In responding	$34	64 hours
Shoulder Strain	Fire Scene	None	None
Shoulder Strain	Fire Scene	$6	2 hours
Shoulder Strain	Training	$152	1 hour
Shoulder Strain	Fire Scene	None	None

Table 10-7. Total Cost for Outside Medical Payments, Injury Leave, and Temporary Disability for an Oklahoma Fire Department

Year	Total Cost
1988-1989	$279,967
1989-1990	$175,244
1990-1991	$255,199
1991-1992	$293,551
1992-1993	$248,766
1993-1994	$347,699

Table 10-8. Average Worker's Compensation Distributions for an Oklahoma Department*

Body Part	Year			
	1990-1991	1991-1992	1992-1993	1993-1994
Arm	NA**	$ 9,334	NA	NA
Ankle	NA	NA	NA	$5,190
Back	$17,798	$50,742	$24,681	$14,419
Ears/Hearing	$10,460	$ 9,000	$ 9,805	$ 3,644
Elbow	NA	NA	$17,344	NA
Eye	NA	NA	NA	$ 2,220
Finger	NA	$ 1,715	NA	NA
Foot	NA	$ 4,094	$13,250	NA
Hand	$ 4,055	NA	$5,190	$ 6,920
Head	NA	NA	NA	$ 3,000
Heart	$17,117	$17,811	NA	$28,592
Knee	$15,927	$7,998	$11,109	$10,965
Leg	$18.320	$17,210	$ 6.467	$14.381
Lower Trunk	NA	NA	$32,113	$17,675
Lungs	$14,280	NA	$13,869	$12,464
Multiple	$32,274	NA	NA	NA
Neck	NA	$18,061	$23,125	NA
Other	$40,664	NA	NA	NA
Shoulder	NA	$24,275	$12,478	$16,650
Upper Trunk	NA	NA	$56,521	$10,961
Whole Body	NA	NA	NA	$13,744

*   Note: injuries occurred in prior years; claims typically occur during the retirement physical exam process
**   NA = not applicable

These costs may be difficult to ascertain ahead of time. In addition, ergonomic issues may be prioritized so that the most improvement can be gained with minimal investment. In general, the greatest increases in costs are associated with the replacement or retrofitting of equipment and especially with station redesign.

A department may not have the appropriate personnel to evaluate ergonomic issues and to design and implement an ergonomics plan. It may be cost-effective for a department to utilize an outside firm specializing in ergonomics. Services can range from short, specific training sessions to the development and management of comprehensive ergonomics and managed care and rehabilitation programs. Locally, ergonomics consulting may be available through the engineering, human factors or human engineering, or industrial engineering departments of colleges or universities. Other sources for local ergonomics consultants may be referrals from engineering departments or the yellow pages of the telephone directory (under "human factors and engineering"). A number of consulting firms offer services nation-wide. Some of these are listed in **Appendix B** (Sources of Information). The Saunders Group is a non-profit group which offers a listing of ergonomics resources (contact: 800-969-4374).

Ergonomics consulting firms can assess a department's needs and then design and implement an appropriate ergonomics program to comply with state and federal regulations. Typically, an outside firm would spend a few days analyzing the stations visiting emergency response scenes, and meeting with relevant personnel. An ergonomics plan would be developed and the department's ergonomics committee or team would be trained. Ergonomic issues could also be prioritized so that those with the greatest impact are addressed first, and limited resources best utilized. Many firms adopt the "train the trainer" approach, so that the outside firm would not necessarily be involved in the actual management of the ergonomics program. However, most firms could assume that function if desired.

Likewise, medical monitoring and injury management and rehabilitation programs can be developed by the outside consultants and, if needed, managed by the group as well. Fees of course can vary a great deal, and depend upon the size of the department, the level of services needed (whether a short training session or extensive program development), and the time involved. Several firms estimated that typical fees for several days of department assessment and the training of the ergonomics team could range from $5,000 to $20,000. Most agreed that $10,000 would be a reasonable "ball-park" estimate for a department to use when initially evaluating the benefit of implementing an ergonomics program.

U. S. Fire Administration Publication FA 141/December 1993, "A Guide to Funding Alternatives for Fire and Emergency Medical Service Departments," provides some guidance for how departments may be able to obtain funding for special programs.

GLOSSARY

Anthropometry - the study of characteristic human physical dimensions, such as distance between anatomical parts

Biomechanics - the use of mechanical principles, such as levers, to describe and analyze the movement and structure of body parts

Blood Pressure - measurement of the pressure exerted by the blood on the walls of the peripheral blood vessels typical at heart level; Systolic pressure indicates the left side off the heart at the end of left ventricle's contraction, while diastolic pressure reflects the resting pressure in the heart between beats

Carpal Tunnel Syndrome - entrapment of the median nerve of the hand and wrist in the tunnel through the carpal bones of the wrist; symptoms include finger numbness and pain upon gripping

Chronobiology - the study of the variations of biological events as a function of time

Circadian Rhythm - describes physical measurements (such as body temperature) that vary periodically during a 24-hour day

Comfort Rating - a subjective indication of the well-being experience during a given set of environmental or task conditions

Compressive Force - a force applied at 90 degrees to a surface; for example, the force experienced by the spine during lifting

Cumulative Trauma Disorder (CTD) - musculoskeletal or neurological symptoms caused by repetitive tasks during which forceful exertions and/or deviations or rotations of the fingers, hands, wrists, elbows or shoulders occur; examples include tendinitis and bursitis

Desynchronization of Rhythms - the disruption of normal physiological rhythms caused by changes in activity or behavior patterns, such as rotating between day, evening, and night shifts

Double Product - an indirect measurement of the work of the heart, calculated by multiplying the heart rate (in beats per minute) by the systolic blood pressure (in millimeters of mercury)

Energy Expenditure - a measurement of the power used during an activity, typically expressed in kilocalories per minute or in millimeters of oxygen per kilogram of body weight per minute; Each liter of oxygen consumed represents 5 kcal of heat production

Ergonomic Design - the use of ergonomic principles in the planning of facilities, tasks, or machines, devices, etc.

Ergonomics - the study of the relationship between human physiological and psychological capabilities and the design of tasks, facilities, and equipment; also known as human factors, human engineering, or human factors engineering

Fatigue - the decrease in physical performance caused by the accumulation of metabolic by-products in tissues and blood as the result excessive or prolonged activity

Heart Rate - the measurement of heart contraction frequency (beat/minute) which is an indicator of work load and other job stresses

Hyperextension of the Spine - position in which the trunk is extended beyond the upright position such that an extreme backward arch is present; often occurs in work performed above shoulder height

Intensity-Duration Relationship - describes the fact that the longer physical work is sustained, the smaller the percentage of maximum work capacity is available for use; in dynamic work, the available aerobic capacity is lessened

Job Design - the distribution of tasks over the work shift; ideally, tasks are arranged to reduce fatigue or the chance of injury or error

Job Restriction - an approach to return injured or chronically ill persons safely to work by designating certain tasks as unsuitable

Kinematics - the biomechanical analysis of movement without consideration of mass or force

Kinetics - in biomechanics, the study of forces affecting human body movement

Lumber Disc - the intervertebral discs between the lumbar vertebrae; under greatest stress during lifting, forward bending, or slumping

Maximum Aerobic Work Capacity - the highest oxygen consumption rate achievable under a given set of conditions; is a measure of cardiovascular fitness level (see physical fitness)

Maximum Grip Span - the greatest distance between thumb and fingers for which a power grip on an object is still possible

Maximum Permissible Limit (MAL)- the NIOSH-recommended upper limit for weight to be handled with two hands in the sagittal plane at 76 cm (30 in) above the floor and at different locations in front of the ankles

Momentum - measurement obtained by multiplying an object's mass by its velocity

Muscoloskeletal - pertaining to muscles, bones, and joints

Myositis - inflammation of a muscle caused by heavy use or repeated use of a muscle with inadequate recovery time

Oblique Grip - grasp in which an object is held in the palm along the base of the thumb with the fingers flexed in different degrees

Overuse Syndromes - see cumulative trauma disorder

Oxygen Consumption - the uptake volume of oxygen across the whole body; used to assess physical workload and/or metabolic heat production

Perceived Exertion - the subjective measurement of effort required for a given task

Phase Shift - the movement of a circadian rhythm's maximum value in time; indicates physiological adaptations to situations such as shift work are occurring

Physical Fitness - indicator of optimum operation of the human body systems; measured in terms of oxygen consumption (maximum) rate

Psychosocial - describes factors with both psychological and social effects, often an issue with shift work

Radial Deviation - movement of the wrist/hand towards the radius bone (thumb side) of the forearm

Recovery Time - periods during which work is light or suspended so that the worker can recover from excessive exertion or environmental conditions

Repetitive-Motion Disorders - see cumulative trauma disorder

Short-Duration Heavy Effort - a period of less than 20 minutes during which physical effort occurs which demands more than 70% of the worker's maximum aerobic capacity

Sleep Satisfaction - psychosocial measure of how rested an individual feels after a rest period; generally lower for persons on the late-night shift who must sleep days

Standard Deviation - the variability of values around the mean of average value

Strain - indicators of stress, such as heart rate, or deformation of a body part in response to increased force per unit area

Stress - physiological, psychological, or environmental effects that can lower performance

Task Analysis - the process by which the physiological and psychological demands of a job are analyzed; over time, the performance required of the worker and the equipment and their interactions are measured and then analyzed

Tendinitis - tendon inflammation, often caused by repetitive, forceful exertions involving rotations around a joint

True Grip Span - the position of the thumb and fingers in a power grip in which near maximum force can be exerted (typically about 2.5 in separation)

Ulnar Deviation - movement of the hand/wrist toward the ulna bode side of the forearm (little finger)

Ventilation - the quantity of air exchanged per unit time between the environment and the air sacs of the lungs

Whole-Body Work - work requiring use of most of the body's muscles (the large muscles of the trunk, arms, shoulders, legs, buttocks); for example, work lower than 30 in above the floor

Work Physiology - the measurement of the body's cardiovascular, respiratory, nervous, and musculoskeletal responses to physical or mental effort, especially as related to industry or one's job

APPENDIX A
ANNOTATED BIBLIOGRAPHY

ANNOTATED BIBLIOGRAPHY

ERGONOMICS - GENERAL

1. Abeysekera, J. D. A., "Some Ergonomics Issues in the Design of Personal Protective Devices," <u>Performance of Protective Clothing Challenges for Developing Protective Clothing for the 1990s</u> Fourth Volume, ASTM STP 1133, ASTM, Philadelphia, PA, 1991, pp. 651-659.

    Use of personal protective devices was evaluated in terms of ergonomic considerations based on surveys of several industrial groups. These surveys indicated that the majority of those exposed to hazards are reluctant to use personal protective devices because of their impact on comfort. Investigations revealed that user-centered designs can reverse these attitudes. The survey also showed manufacturers to be more inclined to adhere to standards, and thus the development of ergonomic standards seems to be a more feasible approach to persuade manufacturers to address user needs.

2. Armstrong, Thomas J. and Ellen A. Lackey, "An Ergonomics Guide to Cumulative Trauma Disorders of the Hand and Wrist." Ergonomics Guide published by AIHA, 1994.

    Summarizes the anatomy and mechanisms of cumulative trauma disorders of the hand and wrist and reviews workplace risk factors.

3. Bert, J. L., <u>Occupational Diseases,</u> British Columbia University, Vancouver, Canada, June 1989.

    This educational module sponsored by NIOSH and the Center for Disease control covers the basic principles un derlying the development of occupational diseases. Included is a brief section on ergonomics which describes ergonomic checklists and common adverse effects caused by the workplace, including cumulative trauma, fatigue, and hearing loss.

4. Collins, B. L. and N. D. Lerner, "Assessment of Fire Safety Symbols," <u>Human Factors Journal,</u> Vol. 24(l), 1982, p.075.

    The understandability of twenty-five internationally proposed symbols was evaluated, and some potentially dangerous ambiguities were revealed.

5. Eastman Kodak Company, <u>Ergonomic Design for People at Work,</u> Volumes 1 and 2, Van Nostrand Reinhold, New York, 1993.

    Uses design examples and case studies to show professionals how to determine which heights, reaches, comfort levels, and visual distance best promote worker comfort and maximum productivity. Provides methods for evaluating human capabilities (i. e., strength and aerobic capabilities) and coordinating and controlling the body's response to increasingly difficult work environments. Includes annotated bibliography, glossary, and list of ergonomics journals.

6. Manning, Harlan T., "Ergonomics: The Science of Profitability," <u>ASTM Standardization News,</u> Vol. 22, February 1994, pp. 32-37.

    Briefly summarizes the history of ergonomics and ergonomic-related injuries and other costs to the workplace, and discusses ergonomic design and proposed ASTM guidelines.

7. Montante, William M., "An Ergonomic Approach to Task Analysis, " <u>Professional Safety,</u> Vol. 39, February, 1994, pp. 18-22.

    The "Trigger/Purpose/Action/Check" process for ergonomic task analysis is described, and a case study involving manual handling is discussed.

8. Phillips, M. D. and R. L. Pepper, "Shipboard Fire-fighting Performance of Females and Males," <u>Human Factors Journal,</u> Vol. 24(3), 1982, p. 277.

    Study results indicated that fire-fighting tasks involving upper torso strength and grip strength challenged female Navy personnel. Experiments included simulations of fire fighting with CO, extinguishers and starting a P-250 pump.

9. Tichauer, E. R. and Howard Gage, "Ergonomic Principles Basic to Hand Tool Design." Ergonomic Guide published by AIHA, 1978.

    Discusses features of hand tools relevant to ergonomic concerns and describes common physiological problems associated with improper hand tool design and usage. Includes a discussion of the ergonomic effects of work gloves.

ERGONOMICS PROGRAMS

1. Burns, David J. and Cherilyn N. Nelson, "A Suggested Strategy to Increase Employee Use of Protective Clothing, " <u>Performance of Protective Clothing: Fourth Volume, ASTM STP 1133,</u> James P. McBriarty and Norman W. Henry, Eds., American Society for Testing and Materials, Philadelphia, 1992, pp. 946-953.

    Discusses the relationship between use of personal protective clothing and individual risk/sensation preference and offers strategies for increasing employee use of protective clothing.

2. Carmean, Gene and Vernon R. Padgett, "A Guide to Evaluating Fire Service Medical-Screening Programs," <u>Fire Engineering,</u> Vol. 142, May 1989, pp. 27-28+.

    Outlines when and how fire service medical-screening examinations should be given; discusses test selection; and addresses legal concerns.

3. Carmean, Gene and Vernon R. Padgett, "Putting Fitness to the Test," Fire Command, Vol. 56, February 1989, pp. 15-19.

   Implications of NFPA 1500 for fire department physicians are discussed. Legal and testing considerations are addressed.

4. Ellam, L. D., et al., "Initial Training as a Stimulus for Optimal Physical Fitness in Firemen", Ergonomics, Vol. 37(5), 1994, pp. 933-941.

   The physical fitness of firemen after the first 18 months in service was compared to the fitness levels before and after initial training. Increases in physical fitness gained during initial training were generally not sustained during service. Recommendations were made for more intense physical training during service.

5. Feare, Tom, "Better Ergonomics: It's the Law!" Modern Materials Handling, Vol. 49, June 1994, pp. 47-49.

   Proposed rules on ergonomics are mentioned. Four case histories are presented to illustrate successful implementation of specific ergonomic design improvements.

6. Keller, David B., "Initiating and Implementing an Ergonomics Program," Am. Ind. Hvg. Assoc. J., Vol. 48, November 1987, pp. 710-713.

   Provides recommendations for putting an ergonomics program together for industry.

7. Landrigan, Phillip J., M. D. and Dean B. Baker, "The Recognition and Control of Occupational Disease," JAMA, vol. 266(5), pp. 676-680.

   A program is outlined for the control of occupational disease based on "preventing exposures in the workplace, premarket toxicity testing of new chemicals and technologies, and astute clinical diagnoses."

8. MacLeod, D., The Ergonomics Edge, Van Nostrand Reinhold, New York, 1993.

   Provides case studies and examples to show how a successful ergonomics program can maximize both safety and profits. Provides views of ergonomic design and application with a careful examination of ergonomics and cumulative trauma with an appraisal of present and future regulatory trends.

9. Tapp, Linda M., "Establishing an Employee Ergonomics Task Force, " Professional Safety, Vol. 39, July 1994, pp. 26-28.

   Basic steps are summarized for setting up an employee ergonomics task force. The importance of including employees is stressed. Sample hazard analysis and product evaluations forms are included.

EQUIPMENT AND PROTECTIVE CLOTHING DESIGN

1. Carlson, Gene P., "Apparatus design for Driver/Operator Safety, " Fire Engineering, Vol. 147, April 1994, pp. 10, 12.

    Discusses design considerations for optimal pump panel location and for increased operator safety.

2. Cerf-Beare , A. , "Environmental Design and Architectural Barriers: An Ergonomic Approach," Man-Environment Systems, Vol. 5(1), January 1975, pp. 49-57.

    Presents "an ergonomic schema" for designers comprised of measurement and evaluation (including use of anthropometrics, surveys, and simulations or mock-ups); consideration of physical and psychological features of the environment; function definition; and safety. Includes an example of ergonomic application to building design.

3. English, William, "Reducing Falling Hazards on Fire Trucks," Professional Safety, Vol. 38, September 1993, pp. 35-38.

    Discusses in general terms the need for further research into accident occurrence and for design improvements to reduce falling hazards on fire trucks, and shows through photographs specific examples of safer truck design (including features such as slip-resistant surfaces, grab bars, etc.).

4. Lundgren, Claes, "Breathing Gear Ergonomics: A Brief Review of a Recent Workshop," Marine Technology Society Journal, Vol. 23, December 1989, pp.34-37.

    Workshop discussions in the following areas are briefly described: respiratory impediments from divers' breathing gear and the underwater environment; modeling and unmanned testing for gear design optimization; gas delivery systems; and psychological effects of breathing gear use. References for specific discussions are included.

5. McConville, John T., "Anthropometric Fit Testing and Evaluation," Performance of Protective Clothing. ASTM STP 900, R. L, Barker and G. C. Coletta, Ed., American Society for Testing and Materials, Philadelphia, 1986, pp. 556-568.

    Guidelines are described for conducting anthropometric fit testing for protective clothing.

6. Pagels, Ted J., "Rebuilding for Safety," Fire Engineering, Vol. 142, May 1989, pp. 36-37.

    A case history is given which details the rebuilding of a pumper to improve safety and upgrade performance. Cost of rebuilding was two-thirds that of purchasing a new apparatus.

7.   Robinette, Kathleen M., "Anthropometric Methods for Improving Protection " Performance of Protective Clothing. ASTM STP 900, R. L. Barker and G. C. Coletta, Eds., American Society for Testing and Materials, Philadelphia, 1986, pp. 569-580.

The development and use of anthropometric sizing system for personal protective equipment is discussed. Advantages for such an approach include better fit and increased safety and productivity.

8.   Rotmann, Manfred F., "Selection and Development of protective Clothing for Fire Fighters -- A Case Study for Users' Assessment of Standards, Tests, and performance Requirements, " Performance of Protective Clothing: Fourth Volume. ASTM STP 1133, James P. McBriarty and Norman W. Henry, Eds., American Society for Testing and Materials, Philadelphia, 1992, pp. 311-321.

Reviews aspects of selection and development of fire fighters' protective clothing, including technical needs, field assessment trials, heat stress, and standards.

9.   Schwind, Gene, "One-Size-Fits-All Inhibits Ergonomic Progress, " Material Handling Engineering. Vol. 44, March 1989, p. 27.

Discusses problems with inadequately sized clothing, tools, and equipment and its impact on industrial worker productivity.

10.   Veghte, James H., "The Physiologic Strain Imposed by Wearing Fully Encapsulated Chemical Protective Clothing," Chemical Protective Clothing Performance in Chemical Emergency Response, ASTM STP 1037, J. L. Perkins and J. 0. Stull, Eds.,American Society for Testing and Materials, Philadelphia, 1989, pp. 51-64.

Physiological responses were determined for firefighters wearing fully encapsulated suits while performing representative tasks under various environmental conditions. Responses depended markedly on suit design. Traditional andfunctional anthropometric dimensions were measured and design considerations discussed.

11.   Veghte, James H., "Physiological Field Evaluation of Hazardous Materials Protective Ensembles," Performance of Protective Clothing: Second Symposium. ASTM STP 989, S. Z. Mansdorf, R. Sager, and A. P. Nielson, Eds., American Society for Testing and Evaluation, Philadelphia, 1988, pp. 461-471.

Three HAZMAT protective ensembles were evaluated under three climatic conditions, Physiologic parameters varied markedly with suit design and with climatic conditions. Suit modifications were suggested for reducing clothing encumbrance and for enhancing work efficiency.

FIRE FIGHTER AND EMS PERSONNEL INJURIES

1.   Gledhill, N. and V. K. Jamnik, "Characterization of the Physical Demands of Firefighting," Can. J. Spt. Sci., Vol 17(3), 1992, pp. 207-213.

       Study included a task analysis of all firefighting operations and physical and physiological characterization of tasks deemed physically demanding. Findings suggested requiring a minimum VO_2max standard of 45ml/kg/min for firefighter applicants.

2.   Guidotti, Tee. L., "Human Factors in Firefighting: Ergonomic-, Cardiopulmonary-, and Psychogenic Stress-Related Issues," Int. Arch. Occup. Environ. Health, Vol 64, 1992, pp. 1-12.

       Reviews studies of cardiopulmonary health, ergonomic, and stress issues for firefighters. Ergonomic issues include energy cost of firefighting activities, heat stress, and encumbrance from personal protective equipment.

3.   Guidotti, Tee L., and Veronica M. Clough, "Occupational Health Concerns of Firefighting," Annual Review of Public Health, Vol. 13, 1992, pp. 151-171.

       Article includes in-depth discussion of specific hazards (thermal, chemical, and psychological), health effects (acute and chronic), ergonomic issues, and 127-entry bibliography for further information.

4.   Heineman, Ellen F., Carl M. Shy, and Harvey Checkoway, "Injuries on the Fireground: Risk Factors for Traumatic Injuries Among Professional Fire Fighters," American Journal of Industrial Medicine, Vol. 15, 1989, pp. 267-282.

       The effects of self-contained breathing apparatus (SCBA) and other risk factors on three types of fire-related injuries (smoke inhalation, burns, and falls) are evaluated in a case-controlled study of a metropolitan fire department.

5.   Karter, M. J., Patterns of Fire Fighter Injuries. 1989-1991, National Fire Protection Association, Fire Analysis & Research Division, Quincy, MA, December 1993.

       Provides an analysis of fire fighter injury data on the Federal Emergency Management Agency's (FEMA) National Fire Incident Reporting System (NFIRS) which determines the factors which contribute to these injuries.

6.   LeCuyer, John, "Where Tools and People Meet," Fire Chief, February 1994, pp. 34-41.

       Discusses general principles of workplace ergonomics and representative problems with some fire fighting equipment, and suggests injury report documentation and ways to reduce injuries by changing practices.

7.   Nuwayhid, Iman A., Walter Stewart, and Jeffrey V. Johnson, "Work Activities and the Onset of First-Time Low Back Pain among New York City Fire Fighters," <u>American Journal of Epidemiology,</u> Vol 137(5), 1993, pp. 539-548.

        Studies the relationship between fire fighting activities and low back pain.

8.   Paley and Tepas, "Fatigue and the Shiftworker: Firefighters Working on a Rotating shift Schedule," <u>Human Factors,</u> Vol. 36(2), June 1994, p. 269.

        Study evaluates the effects of shift on time-of-day interactions in firefighters' sleep length, sleepiness, and mood. The night shift had the greatest effect, and effects were not mitigated by the end of the two-week shift period.

9.   Pine, John C., "Firefighter Injuries and Illnesses in Louisiana," <u>Speaking: of Fire,</u> Spring, 1992.

        Statistical information is presented for occupational injuries and illnesses for fire fighters in Louisiana from 1985 to 1990.

FIRE FIGHTER PSYCHOLOGICAL DISORDERS

1.   Boxer, Peter A. and Deanna Wild, "Psychological Distress and Alcohol Use Among Fire Fighters," <u>Scand. J. Environ. Health,</u> Vol. 19, 1993, pp. 121-5.

        Study findings suggest that more than one-third of fire fighters experience significant psychological distress. 29 % had possible problems with alcohol use. However, logical regression analysis did not show a relationship between measures of psychological distress and alcohol use and the 10 most highly ranked work stressors.

2.   Guidotti, Tee. L., "Human Factors in Firefighting: Ergonomic-, Cardiopulmonary-, and Psychogenic Stress-Related Issues," <u>Int. Arch. Occup Environ. Health,</u> Vol 64, 1992, pp. 1-12.

        Reviews studies of cardiopulmonary health, ergonomic, and stress issues for firefighters. Psychological issues include risking of personal security, stress related to episodes of high danger or tragedy, and post traumatic stress disorder.

3.   Reichel, Doug, Ph.D. Dissertation on Fire Fighter Burnout. University of Florida, To be completed in 1994-1995.

        Fire fighter burnout will be analyzed in terms of 6.5 factors.

4.   Shearer, Robert W., Occupational Stress in the Fire Service," <u>Professional Safety,</u> Vol. 34, April 1989, pp. 22-25.

        The three physical stages of stress are outlined, Sources of fire fighter occupational stress are discussed, as are its effects on individual health, substance abuse, personal/family life, and burnout. Recommendations to reduce stress include job rotation, physical fitness programs, additional training sessions addressing stress, and psychological services.

DEXTERITY AND RANGE-OF-MOTION IMPAIRMENT

1. Adams, Paul Stuart, <u>The Effects of Protective Clothing on Worker Performance: a Study of Size and Fabric Weight Effects on Range-of-Motion.</u> PhD dissertation, University of Michigan, 1993.

    The relationships between garment properties and worker performance were studied, and a "Garment Impediment Index" model was proposed. Range-of-motion (ROM) was evaluated via three methods: a universal goniometer, a Leighton Flexometer, and an electrogonimeter. In addition, two studies addressed the effect of garment weight on arm movement speed and the ROM effects of subject anthropometry and garment dimensions.

2. Ashdown, Susan P. and Susan M. Watkins, "Movement Analysis as the Basis for the Development and Evaluation of a Protective Coverall Design for Asbestos Abatement," <u>Performance of Protective Clothing: Fourth Volume, ASTM STP 1133,</u> James P. McBriarty and Norman W. Henry, Eds., American Society for Testing and Materials, Philadelphia, PA, pp. 660-674.

    The development of a movement analysis test for identifing movement problems associated with disposable coveralls used by asbestos abatement workers is discussed, and test results used to create a new coverall design.

3. Bensel, C.K., <u>The Effects of Various Thicknesses of Chemical Protective Gloves on Manual Dexterity.</u> Ergonomics, June 1993, 36(6): 687-96.

    Study results indicate that use of the thinnest glove material compatible with working hazards and practice with the chosen gloves results in relatively efficient manual performance.

4. Veghte, J.H., "Functional Integration of Fire Fighters' Protective Clothing," <u>Performance of Protective Clothing. ASTM STP 900,</u> R. L. Barker and G. C. Coletta, Eds., American Society for Testing and Materials, Philadelphia, 1986, pp. 487-496.

    Approaches for the integration of firefighter's protective clothing were examined in combining coats, pants, helmet, boots, gloves, and breathing apparatus. Functional integration of characteristics such as resistance to cuts and punctures, mobility, waterproofness, fit, and durability were compared with performance features such as thermal protective performance (TPP) rating analyses of several hundred protective clothing combinations.

HEAT STRESS

1.   Blankinship, Julian A. and William P. Behnke, "Turnout Coats Components Tested," Fire Chief Magazine, Vol. 22(12), December 1978, pp. 36-38.

      The physiological effects on fire fighters of turnout coat shells, vapor barriers, and liners were evaluated.

2.   Davis, P. O., "Physiological Aspects of Fire Fighting," Fire Technology, Vol. 23(4), November 1987, pp. 280-291.

      The environmental stressors to which fire fighters are exposed (temperature, noise, breathing atmosphere, and protective gear) are discussed and related to the importance of evaluating fire fighter cardiovascular and orthopedic fitness to minimize injury risk.

3.   Dukes-Dobos, F. N., Reischl, U., Buller, K., Thomas, N. T., and Bernard, T. E., "Assessment of Ventilation of Firefighter Protective Clothing," Performance of Protective Clothing: Fourth Volume, ASTM STP 1133, J. P. McBriarty and N. W. Henry, Eds., American Society for Testing and Materials, Philadelphia, 1992, pp. 629-633.

      This study used a gas dilution technique to assess ventilation of firefighter protective ensembles. The cuffs of the sleeves and pants, and the collar and front closure of the turnout coat were opened and closed to determine the effect on ventilation. In addition, the effect of using a belt or suspenders to hold the pants was examined. The gas dilution technique provided quantitative data on the increase in garment ventilation due to various combination of the openi ngs and suspension of the pants. The greatest effect resulted from opening the collar and pant cuffs along with the use of suspenders.

4.   Haver, C. H. M., "Protective Clothing for Firefighting Personnel," Commission of the European Communities, Fires in Buildings, 1984, pp. 390-399.

      Discusses the minimum requirements for protective clothing for fire fighters, including physiological and ergonomic aspects. Flame testing, TPP testing, and conducted heat testing are briefly described.

5.   Holmer, I., "Thermal Properties of Protective Clothing and Prediction of Physiological Strain," Performance of Protective Clothing: Second Symposium, ASTM STP 989, S. Z. Mansdorf, R. Sager, and A. P. Nielsen, Eds., American Society for Testing and Materials, Philadelphia, 1988, pp. 82-86.

      Thermal insulation (I_T) and evaporative resistance (R_T) of garments and p rotective clothing systems were measured using indirect calorimetry on subjects exercising in a climatic chamber. In different series of wear trials, the physiological responses associated with wearing these garments during 60 min of treadmill walking in hot environments (35 to 45°C, 30 to 40% relative humidity) were evaluated for young, male, healthy subjects. Obtained values for I_T and R_t were used in a physiological model to predict rectal temperature response to the same conditions as for the wear trials in order *to compare predicted and measured rectal temperatures. The results showed that a combination of clothing heat exchange measurements on subjects and simple physiological modeling may be a useful technique for assessment of clothing thermal function.*

6. Ohnaka T., Y. Tochihara, and T. Muramatsu, "Physiological Strains in Hot-Humid Conditions while Wearing Disposable Protective Clothing Commonly Used by the Asbestos Removal Industry," Ergonomics, Vol. 36(10), October 1993, pp. 1241-1250.

The thermal stress caused by working in hot-humid conditions with disposable protective clothing was greatly minimized by rest in a cool environment between work periods.

7. Olesen, B. W., and F. N. Dukes-Dobos, "International Standards for Assessing the Effect of Clothing on Heat Tolerance and Comfort, " Performance of Protective Clothing: Second Symposium, ASTM STP 989, S. Z. Mansdorf, R. Sager, and A. P. Nielson, Eds., American Society for Testing and Materials, Philadelphia, 1988, pp. 17-30.

Three standards, developed by the International Organization for Standardization (ISO) Working Group for Thermal Environments for assessing thermal load of workers, are discussed in terms of clothing effect on workers' heat tolerance and comfort. Standard limitations are also discussed.

8. Shults, R. A., G. P. Noonan, N. L. Turner, R. M. Ronk, "Investigation of a Heat Stress-Related Death of a Fire Fighter," Fire Technology, Vol. 28, November 1992, pp. 317-331.

The investigation by NIOSH of a fire fighter's death by heat-stroke during response to a brush fire is detailed. Recommendations were made for incident command and safety procedures, medical monitoring, fire scene rehabilitation of fire fighters, and rehydration schedules.

9. White, M. K., T. K. Hodous, and M. Vercruyssen, "Effects of Thermal Environment and Chemical Protective Clothing on Work Tolerance, Physiological Responses, and Subjective Ratings," Ergonomics, April 1991, 34(4): 445-57.

Physiological and subjective responses were studied for men wearing SCBA with either light work clothing or chemical protective clothing in each of three thermal environments. Wearing the chemical protective suit resulted in greater measured and perceived stress.

10. White, Mary Kay and Thomas Houdous, "Physiological Responses to the Wearing of Fire Fighter's Turnout Gear with Neoprene and Gore-tex[R] Barrier Liners," American Industrial Hygiene Association Journal, Vol. 49(10), 1988, pp. 523-530.

Study results indicated that the physiological benefits generally associated with vapor permeable garments such as Gore-tex[R] liners are minimized when the garments are worn with fire fighter's turnout gear during sustained moderate to heavy work in a warm environment.

11.   White, M. K., M. Vercruyssen, and T. K. Hodous, "Work Tolerance and Subjective Responses to Wearing Protective Clothing and Respirators During Physical Work," Ergonomics. Sept 1989, 32(9): 111 1-23.

Work tolerance and subjective responses were studied for individuals performing two levels of work while wearing one of four protective ensembles. The ensembles included light work clothing; light work clothing with SCBA; firefighter turnout gear with SCBA; and chemical protective clothing with SCBA. Wearing protective clothing and SCBA caused significant thermoregulatory and cardiovascular stress, with turnout gear having the greatest impact.

FITNESS PROGRAMS

1.   Barnard, R. James, and Donald F. Anthony, "Effect of Health Maintenance Programs on Los Angeles City Firefighters," Journal of Occupational Medicine. Vol. 22 (10), October, 1980, pp. 667-669.

Evaluation of injury and fitness data for Los Angeles fire fighters from 1972 to 1978 indicated the success of the health maintenance program in increasing the fitness level and reducing the risk factors associated with atherosclerotic heart disease.

2.   Ben-Ezra, Victor and Richard Verstraete, "Stair Climbing: An Alternative Exercise Modality for Firefighters, " Journal of Occupational Medicine. Vol. 30 (2), February 1988, pp. 103-105.

The maximal cardiorespiratory responses of fire fighters during stair climbing was compared to responses during maximal graded treadmill exercise. Results suggested that a stair-climbing device would be more appropriate for testing of fire fighters because stair climbing is a task-specific activity for fire fighting, and that the difference in oxygen consumption between treadmill and stair-climbing exercise should be considered when fitness levels for fire fighters are recommended.

3.   Brown, A., J. E. Cotes, I. L. Mortimore, and J. W. Reed, "An Exercise Program for Firemen. " Ergonomics. Vol. 25 (9), 1982, pp. 793-800.

Physical fitness of 30 fireman was assessed using physiological indices obtained during cycle ergometry. Fitness levels were lower than those for mine rescue brigadesmen, heavy-industry factory workers, and U. S. volunteer fire fighter recruits of comparable age. Implementation of a physical fitness training regimen resulted in increases in maximal oxygen uptake, maximal stroke volume and other indices.

4.   Cady, Lee D., David P. Bischoff, Eugene R. O'Connell, Phillip C. Thomas, and John H. Allan, "Strength and Fitness and Subsequent Back Injuries in Firefighters," Journal Of Occupational Medicine. Vol. 21 (4), April 1979, pp. 269-272.

The relationship between five strength and fitness measurements and subsequent back injuries was investigated for 1652 Los Angeles fire fighters for the period 1971 to 1974. Results indicated that physical fitness and conditioning are preventive of back injuries.

5.   Cady, Lee D., Phillip C. Thomas, and Robert J. Karwasky, "Program for Increasing Health and Physical Fitness of Fire Fighters," Journal Of Occupational Medicine, Vol. 27 (2), February 1985, pp. 110-114.

Implementation of a health and fitness program for Los Angeles fire fighters resulted in increases in physical work capacity and a decrease in the percentage of habitual smokers. High levels of physical work capacity, strength, and flexibility were inversely related to workers' compensation costs. Myocardial infarctions occurred 2.6 times more frequently in fire fighters with low levels of physical work capacity..

6.   Gledhill, N. and V. K. Jamnik, "Characterization of the Physical Demands of Firefighting," Canadian Journal of Sport Sciences, Vol. 17 (3), 1992, pp. 207-213.

The most demanding, essential firefighting operations were identified by task analysis. The physical demands of these tasks were characterized by physical and physiological evaluation. Results indicated that fitness screening should assess muscular strength and endurance in both upper and lower body, as well as agility, manual dexterity, dynamic balance and flexibility. To meet the demands of fire fighting, a VO_2max standard of 45 ml/kg/min is recommended.

7.   Hilyer, J. C., K. C. Brown, A. T. Sirles, and L. Peoples, "A Flexibility Intervention to Reduce Incidence and Severity of Joint Injuries among Municipal Fire Fighters," J. Occupational Medicine, Vol. 32(7), July 1990, pp. 631-637.

A study of 469 firefighters examined the effect of flexibility training on the incidence and severity of joint injuries. Both flexibility measures and costs (lost time and medical care costs) were investigated in this study. Significant differences were found in flexibility scores of the experimental and control subjects with overall flexibility increases in the experimental group. Although the incidence of injury was not significantly different for experimental and control groups, injuries sustained by the experimental group resulted in significantly less lost time costs. Findings indicate that the flexibility training program had a beneficial effect on reducing the severity and costs of joint injuries in this fire fighter population.

8.   Davis, Paul O., Charles O. Dotson, and D. Laine Santa Maria, "Relationship Between Simulated Fire Fighting Tasks and Physical performance Measures," Medicine and Science in Sports and Exercise, Vol. 14 (1), 1982, pp, 65-71.

Twenty-six performance variables were assessed for 100 professional fire fighters during five sequentially performed fire fighting tasks. Analysis revealed that two physical work capacity and resistance to fatigue accounted for the fractionated time and heart rate data. Certain physical performance variables best predicted each factor. Regression equations for predicting fire fighter physical performance were generated, and a physical performance rating scale was developed.

9. Nutter, P., "Aerobic Exercise in the Treatment and Prevention of Low Back Pain," Occupational Medicine: State-of-the-Art Reviews, Vol. 3(1), January-March 1988, Philadelphia, Hanley & Belfus, pp. 137-145.

Recommends aerobic exercise as part of the treatment for virtually all causes of low back pain. Indicates that the major objective is to keep the body moving.

MUSCULOSKELETAL DISORDERS

1. Conrad, Karen M., George I Balch, Paul A. Reichelt, Sharon Muran, and Kyoung Oh, "Musculoskeletal Injuries in the Fire Service," AAOHN Journal, Vol. 42 (12), Dec. 1994, pp. 572-581.

Focus groups were used to investigate fire chiefs and fire fighters perceptions of the factors contributing to musculoskeletal injury of firefighters and to determine acceptable protective strategies. Fire chiefs and fire fighters agreed on many issues related to musculoskeletal injuries, but differed on the relative contributions to musculoskeletal injuries of person, workplace, and uncontrollable external environmental fa ctors.

2. Garg, Arun, "Occupational Biomechanics and Low Back Pain," Occupational Medicine: State-of-the-Art Reviews, Vol. 7(4), October-December 1992, Philadelphia, Hanley & Belfus, pp. 609-628.

Provides overview of lower back pain and discusses contributations of ergonomic risk factors.

3. Garg, Arun and J. S. Moore, "Prevention Strategies and the Low Back in Industry," Occupational Medicine: State-of-the-Art Reviews, Vol. 7(4), October-December 1992, Philadelphia, Hanley & Belfus, pp. 629-639.

Review the effectiveness of three different approaches to reducing the incidence of lower back pain; (1) reducing number of painful episodes, (2) allowing worker to stay at work longer, and (3) permitting the worker with lower-back disability to return to work sooner. Researchers conclude that only job-specific strength testing and ergonomic job design are partially effective in preventing low back injuries. Ergonomic job design has greatest potential for reducing the incidence of low back injuries.

4. Grew, N. D. and G. Deane, "The Physical Effect of Lumbar Spinal Supports," Prosthetics and Orthotics International, Vol . 6, 1982, pp. 79-87.

Study investigated the effects of lumbar supports using two groups of subjects. The results showed that the longer wearing of lumbar supports increase physical dependence on the devices.

5. Guignard, J. C., "Evaluation of Exposures to Vibrations," Patty's Industrial Hygiene and Toxicology Volume II, ed. by L. J. Craley and L. V. Craley, John Wiley and Sons, New York, 1979, pp. 480-485.

One of the original works which discusses work disorders sue to repeated vibration including vehicular vibration. Provides a diagnosis of different injuries and recommends solutions for avoiding injuries.

6.   Hedlund, G., "A Study Comparing the Amount of Force Exerted on Musculoskeletal Tissues When Using a Short Versus a Long Hand Tool," <u>Vimmerby,</u> Sweden, Farmer's Health Society, 1992.

Examines differences in worker injury based on the length of shovels, rakes, and other hand farm implements.

7.   Keyserling, W. M., D. S. Stetson, B. A. Silverstein, "A Checklist for Evaluating Ergonomic Risk Factors Associated with Upper Extremity Cumulative Trauma Disorders," <u>Ergonomics,</u> Vol. 36, July 1993, pp. 807-831.

A two-page checklist was developed for identifying tasks likely to present ergonomic risks for upper extremity cumulative disorders. The efficacy of the checklist as a sensitive screening tool was established by comparing checklist results to results from expert evaluations.

8.   Khalil, T. M., <u>Ergonomics in Back Pain,</u> Van Nostrand Reinhold, New York, 1993.

Examines the widespread problem of low back pain from its basic anatomic and physiological foundations to advance applications of ergonomics and bioengineering in both the prevention and rehabilitation of lower-back injury. Focuses on key workplace issues, including patient evaluation, functional capacity assessment, workplace design, body mechanics education, functional electrical stimulation, and biofeedback.

9.   Lusa, S., V. Louhevaara, J. Smolander, K. Kinnunen, 0. Korhonen, and J. Soukainen, "Biomechanical Evaluation of Heavy Tool-Handling in Two Age Groups of Firemen," <u>Ergonomics,</u> Vol. 34 (12), 1991, pp. 1429-1432.

This study evaluated the biomechanical load factors in a simulated rescue-clearing task for firefighters in two age groups, To simulate clearing of passages through the ceiling, one of the most physically demanding fire-fighting and rescue tasks, subjects lifted a 9 kg power saw from the floor to a bar 211 cm above floor level. Results indicated that this lifting produced a high load on the musculoskeletal system. The dynamic compressive force at the L5/SI disc did not vary significantly with age, nor did the peak torques for the back and knee extensions. Younger subjects did exhibit higher movement speed in the knee extension.

9.   McGill, S. M., R. W. Norman, and M. T. Sharratt, "The Effect of an Abdominal Belt on Trunk Muscle Activity and Intra-Abdominal Pressure During Squat Lifts," <u>Ergonomics,</u> Vol. 33(2), pp. 147-160.

A study was conducted to determine whether abdominal belts such as those used by industrial workers reduced trunk muscle activity. Study involved control and test groups simulating various lifting exercises. Intra-abdominal pressures were measured for each test subject. Results showed that individual perceptions of support were noted, muscle activity and intra-abdominal pressure measurements made it difficult to prescribe the use of abdominal belts for workers.

10.   Marras, W. S., et al, "Dynamic Measures of Low Back Performance." Ergonomics Guide published by AIHA, 1993.

Discusses different techniques for measuring musculoskeletal motor performance of the lower back.

7.   McAtamney, L., E. N. Corlett, "Ergonomic Workplace Assessment in a Health Care Context," Ergonomics, Vol. 35, September 1992, pp. 965-978.

A procedure is detailed for the ergonomic evaluation of health care workplaces. The Appendix includes a detailed questionnaire that covers tasks, equipment, job design, and organization.

8.   Morrissey, S. J., M. M. Ayoub, Charles L. Burford, and J. D. Ramsey, "Physiological Effects of Stoopwalking and Crawling in Trained Male and Female Subjects," in Ergonomics - Human Factors in Mining, Proceedings from Bureau of Mines Technology Transfer Seminars, 1981, Pittsburgh, PA, pp. 17-32.

The difference in male and female physiological responses to stoopwalking and crawling is evaluated and discussed, and recommendations are made for reducing strain on females by additional training and by permitting self-paced work-rates.

9.   Schneider, S. and P. Susi, "Ergonomics and Construction: A Review of Potential Hazards in New Construction," American Industrial Hygiene Journal, Vol. 55(7), July 1994, pp. 635-649.

Reviews potential ergonomic hazards in new construction work by summarizing findings from published literature and from a 15-month investigation of health hazards in a new construction site in suburban Washington, D.C. The review structure follows the sequence of events in the construction process. Available ergonomic solutions are included.

10.   Wos, W., J. Lindberg, R. Jakus, and S. Norlander, "Evaluation of Building Impact Loading in overhead Work using a Bolt Pistol Support," Ergonomics, Sept. 1992, Vol. 9, pp. 1069-1079.

Describes the effects of using a support rig to assist workers doing overhead work with a pneumatic bolt device. May be applicable to firefighters pulling ceilings in overhaul operations.

OTHER FIRE FIGHTING HAZARDS

1. Agnew, Jaqueline, Melissa McDiarmid, A. Lees, S. J. Peter, and Richard Duffy, "Reproductive Hazards of Fire Fighting I. Non-Chemical Hazards," <u>American Journal of Industrial Medicine,</u> Vol. 19, 1991, pp. 433-445.

    Evidence suggests that heat, noise, and physical exertion may adversely affect reproductive health.

2. Guidotti, Tee L., and Veronica M. Clough, "Occupational Health Concerns of Firefighting," <u>Annual Review of Public Health,</u> Vol. 13, 1992, pp. 151-171.

    Article includes in-depth discussion of specific hazards (thermal, chemical, and psychological), health effects (acute and chronic), ergonomic issues, and 127-entry bibliography for further information.

3. Lusa, Sirpa, Veikko Louhevaara, Juhani Smolander, Mika Kivimaki, and Olli Korhonen, "Physiological Responses of Firefighting Students During Simulated Smoke-Diving in the Heat," <u>Am. Ind. Hyg. Assoc. J.,</u> Vol. 54(5), 1993, pp. 228-231.

    Fire fighting students wearing full protective gear performed smoke-diving (entry into smoke-filled room) during a simulated shipboard fire. Physiological measurements indicated that this activity was very physically demanding for even young and fit students. Previous experience and ability to tolerate stress did not affect heart rate or estimated oxygen consumption during the tests.

APPENDIX B

SOURCES FOR ADDITIONAL INFORMATION

SOURCES FOR ADDITIONAL INFORMATION

One of the first tasks in establishing an ergonomic program should be familiarization with the literature of ergonomics and human factors engineering. The objective is to identify resource and background material that will be instrumental in the development of an up-to-date ergonomics program, and helpful in training personnel. The entire area of industrial ergonomics and safety is presently growing rapidly as is the literature being produced. An annotated bibliography, up to date as of this writing, is included in this manual within Appendix G. General sources of information are listed below.

Textbooks & Scientific Journals

So far as is known, no textbook to date deals specifically with the ergonomics of fire fighting or the EMS but there is a great deal of relevant material. We recommend that one team member be assigned the task of either writing to various publishing houses or visiting the local library. One publishing house that specializes in ergonomic topics is:

> Taylor & Francis, Inc.
> 1900 Frost Road, Suite 101
> Bristol, PA 19007

One of the most comprehensive texts on ergonomics is "Ergonomic Design for People at Work," Volumes I and II. Volume I describes ergonomic principles for work place design, tool selection, and other ergonomic controls. Volume II provides detailed background information on ergonomic risk factors. Both volumes are extremely useful for gaining an understanding about any ergonomic principle, disorder, or solution. These books are available from the Eastman Kodak Company, Ergonomics Group, Health and Environmental Laboratories through Van Nostrand, New York.

All but the smallest of public libraries now have computerized holdings. Following the user instructions at your library, search for ERGONOMICS or HUMAN FACTORS. A list of holdings will appear on the monitor. If the library you use is not computerized, ask the librarian. He/she will direct you to the appropriate area where you may browse through the texts at your leisure. For example, in addition to the traditional journals such as *Ergonomics* and the *Journal of the Human Facto rs and Ergonomic s Society* (HFES) there are a number of journals now being published which tend to be more specially oriented toward applied aspects, e.g., *Applied Ergonomics* (Great Britain), *International Journal of Industrial Ergonomics* (Elsevier Science Publishers, Netherlands). *Ergonomics Abstracts* (Taylor and Francis Limited, London), which has been published quarterly since 1968 by the Ergonomics Information Analysis Centre at the University of Birmingham, annotates over 3000 international publications annually. The HFES in this country also publishes a *Cumulative Index to Human Factors*. Within the United States, HFES can be reached at the following address:

> Human Factors and Ergonomics Society (HFES)
> P. 0. Box 1369, Santa Monica, CA 90406 (310-394-1811)

Trade Journals and Magazines

In addition to the scientific literature there exist quite a large number of trade magazines and journals published by both industry and fire fighting/EMS organizations. These, as well as the more traditional sources, such as the medical literature dealing with emergency medicine, industrial medicine, trauma, and injury prevention should be examined for articles dealing with ergonomics and/or human factors. Examples include:

1. American Industrial Hygiene Association Journal
   2700 Prosperity Avenue
   Suite 250
   Fairfax, VA 22031

2. American Journal of Industrial Medicine

3. ASTM Standardization News
   ASTM
   1916 Race Street
   Philadelphia, PA 19103

4. Fire Chief
   Argus Business
   35 East Wacker Drive
   Suite 700
   Chicago, IL 60601

5. Fire Engineering
   P. 0. Box 1289
   Tulsa, OK 74101

6. NFPA Journal
   NFPA
   1 Batterymarch Park
   Quincy, MA 02269

Organizations

Several groups are involved in writing ergonomic standards or providing guidelines to their respective memberships. Among these are:

1. American Industrial Hygiene Association (AIHA)
   2700 Prosperity Avenue
   Suite 250
   Fairfax, VA 22031

2. American National Standards Institute (ANSI)
   1430 Broadway

New York, NY 10018

3.   International Organization for Standardization (ISO)
        Case Postale 56
        CH-1211 Geneve 20
        Switzerland

4.   National Fire Protection Association (NFPA)
        1 Batterymarch Park
        Quincy, MA 02269
        (6 17-770-3000)

Government Publications

The National Institute of Occupational Health and Safety (NIOSH) which sponsors and performs research in this area is also a good source of information. A list of recommended NIOSH and other government agency publications follows:

1.   *Work Practices Guide for Manual Lifting*
        DHHS (NIOSH) Publication No. 81-122, March 1981
        U.S. Department of Health and Human Services
        Public Health Service, National Institute for Occupational Health and Safety
        Division of Biomedical and Behavioral Science
        Cincinnati, Ohio 45226

2.   *Guidelines for Controlling Hazardous Energy During Maintenance and Servicing*
        DHHS (NIOSH) Publication No. 83-125, Sept, 1983
        U.S. Department of Health and Human Services
        Public Health Service, National Institute for Occupational Health and Safety
        Division of Safety and Health
        Morgantown, West Virginia, 26505

3.   *A Guide to Safety in Confined Spaces*
        Petit, T. and H. Linn
        DHHS (NIOSH) Publication No, 87-113, July, 1987
        U. S. Department of Health and Human Services
        Public Health Service, National Institute for Occupational Health and Safety
        Division of Safety and Health
        Morgantown, West Virginia, 26505

4.   *Preventing Illness and Injury in the Workplace*
        Office of Technology Assessment
        OTA-H-256, April 1985
        Washington, D.C.: U.S. Congress

5.   *Recordkeeping Requirements under the Occupational Safety and Health Act of 1970*

Publication No. 19770-251-096, 1978
U. S. Government Printing Office
Washington, D.C.

6.   *Preemployment Strength Testing*
        Chaffin, D.B., G.D. Herrin, W.M. Keyserling, and J.A. Foulke
        DHEW (NIOSH) Publication No. 77-163, May, 1977
        U.S. Department of Health and Human Services
        Public Health Service, National Institute for Occupational Health and Safety
        Physiology and Ergonomics Branch
        Cincinnati, Ohio 45226

7.   *Scientific Support Documentation for the Revised 1991 NIOSH Lifting Equation*
        National Technical Information Service
        PB91-226274,1991
        U.S. Department of Commerce
        Springfield, Virginia

8.   *An Improved Method for Monitoring Heat Stress Levels in the Workplace*
        Mutchler, J.E., D.D. Malzahn, J.L. Vecchio, and R.D. Soule
        HEW (NIOSH) Publication No. 75-161, May 1975
        U.S. Department of Health and Human Services
        Public Health Service, National Institute for Occupational Health and Safety
        Division of Laboratories and Criteria Development
        Cincinnati, Ohio 45202

9.   *Ergonomics Program Management Guideline for Meatpacking*
        Occupational Health and Safety Administration
        OSHA 3123, 1990
        U.S. Department of Labor

10.   *Participatory Ergonomic Intervention in Meatpacking Plants*
        NIOSH Publication No. 94-124, 1994
        U.S. Department of Health and Human Services
        Public Health Service, National Institute for Occupational Health and Safety
        Physiology and Ergonomics Branch
        Cincinnati, Ohio 45226

11.   *Ergonomic Safety and Health Management; Proposed Rule*
        Federal Register, Vol. 57, No 149
        29 CFR Part 1910, August 3, 1992
        Department of Labor, Occupational Safety and Health Administration
        Washington, D.C.

12.   *Occupational Exposure to Hot Environments: Revised Criteria 1986*
        U. S. Department of Health and Human Services
        Public Health Service; Center for Disease Control

National Institute for Occupational Safety and Health
April 1986

The U. S. Fire Administration of the Federal Emergency Management Agency has conducted a number of studies which provide useful information in the area of ergonomics, particularly heat stress and protective clothing:

1. *Qualitatively Evaluating the Comfort, Fit, Function, and Integrity of Chemical Protective Suit Ensembles*
    FA-107/September 1991

2. *Field Evaluation of Chemical Protective Suits*
    FA-108/September 1991

3. *Physiologic Field Evaluation of Hazardous Materials Protective Ensembles*
    FA-109/September 1991

4. *Emergency Incident Rehabilitation*
    FA-114/July 1992

5. *A Handbook on Women in Fire Fighting*
    FA-128/January 1993

6. *Protective Clothing and Equipment Needs of Emergency Responders for Urban Search and Rescue Missions*
    FA-136/September 1993

7. *Minimum Standards on Structural Fire Fightering Protective Clothing and Equipment: A Guide for Fire Service Education and Procurement*
    FA-137/September 1993

8. *A Guide to Funding for Fire Emergency Medical Service Departments*
    FA 141 /December 1993

9. *EMS Safety: Techniques and Applications, 1994*

10. *Fire Department Hearing Conservation Program Manual, 1991*

11. *Fire Service Fitness Programs, 1977*

12. *Physical Fitness Coordinators Manual for Fire Departments, 1990*

13. *Stress Management Model Program for Maintaining Fire Fighter Well-Being, 1990*

Computer Searches and Annotated Bibliographies

Computerized literature searches which access a great variety of databases can be obtained through the government (National Technical Information Service, NTIS), almost any college or university library, and most county-level public libraries. The selection of the key search words used is vital, since, if the key words are too general, one can be flooded with a pile of useless printout. For example, in this case, use of the key word *ergonomics* alone would currently produce such a result, so it is important to be more specific. The charge for this service is usually nominal and depends, to some extent, on the period of time covered and the complexity of the search ordered. Annotation can be obtained with most computer searches. The quality of both the information presented and the printouts supplied by search organizations are often quite poor. Depending upon the availability of funds in your department you can obtain copies of the more promising articles via interlibrary loans. A good annotated bibliography of useful ergonomics textbooks is given in *Ergonomic Design for People at Work* compiled by the ergonomics group at Eastman Kodak Company (Van Nostrand Reinhold, vol. 2, 1986). The NIOSH also periodically publishes an annotated bibliography related to a specific area of occupational health or safety. To obtain copies of such reports, including those listed above you can write to:

U.S. DEPARTMENT OF HEALTH AND HUMAN SERVICES
Public Health Service
Centers for Disease Control
National Institute for Occupational Health and Safety
Physiology and Ergonomics Branch
Cincinnati, OH 45226

Videos and CD-ROMS

If your department has access to either a VCR or a multimedia computer equipped with a compact CD-ROM, you will be able to access valuable training aids. The best source for information about the availability of such training aids may be the NIOSH or the HFES.

APPENDIX C

SAMPLE ERGONOMICS PROGRAM

SAMPLE ERGONOMIC PROGRAM

Introduction

This is a sample ergonomics program and can be modified for a specific fire or emergency medical services department.

Purpose and Summary

The purpose of this ergonomic management program is to define lines of responsibilities and a set of guidelines that can integrate ergonomic control procedures into all aspects of the department. The goal of this procedure is to help in eliminating injuries associated with repetitive motion and manual material handling. This procedure will also provide direction toward meeting any applicable regulatory health and safety standards related to the fire and emergency medical services department ergonomics issues.

Summary

The major occupational risk factors for Cumulative Trauma Disorders (CTD) include repetitiveness, force, posture, and the use of certain types of hand tools. These risk factors should be addressed as part of the total work environment that will include workplace and tool design, lighting, temperature, the nature of the tasks, and the way the fire fighter or emergency medical technician (EMT) performs their assigned work.

Upper extremity cumulative trauma disorders are disorders caused or aggravated by repeated exertion. Consequently, CTDs are often called repetitive motion disorders. Occupationally related CTDs affect the muscle and skeletal systems or the nerves in the extremities. The upper extremities include the fingers, hands, wrists, forearms, elbows, arms, and shoulders. Injuries of the upper extremities can be classified as acute or chronic. Acute injuries usually occur suddenly during a single event or an accident. Chronic injuries usually develop slowly over a period and may involve many events. Usually, CTDs of the upper extremities are chronic injuries.

Fire fighters and emergency medical technicians will be subject to other ergonomic hazards, including heat stress, cold stress, noise stress, and visual stress. These hazards can also contribute to personnel injuries and make emergency operations difficult. Special personal protective equipment (PPE) or other equipment is often needed to limit exposure to these hazards.

This ergonomics program addresses workplace injuries, increases awareness among all personnel, and provides the appropriate management training. Ergonomics awareness among all levels of the department must be established to perform the appropriate activities such as: station planning and design; response tactics and safety, injury cost tracking and control; equipment and tool selection and purchasing; and training.

Definitions

1. *Anthropometric* - This is the study of people in terms of their physical dimensions. It includes the methods of measuring human body size and form.

2. *Biomechanics* - The application of mechanical principles, such as levers and forces, to the analysis of body part structure, and movement that includes studies of range, strength, endurance, speed of movement, and mechanical response to such physical forces as acceleration and vibration.

3. *Cumulative Trauma Disorders (CTDs)* - Normally, CTDs are associated with repeated or sustained activities that expose tissues to mechanical or overexertion stress.

4. *Dynamic Work* - Dynamic work is the opposite of static work. In dynamic work, the muscles are always changing position. They are continually contracting and relaxing.

5. *Ergonomics* - A multidisciplinary science dealing with the interactions between people and the total work environment to achieve optimum adjustment with the goal of reducing unnecessary physiological and psychological stress and the resultant strain.

6. *Ergonomic Exposure* - A workplace condition that poses a biomechanical stress or stressors on workers. Workplace conditions include, but are not limited to, faulty workstation layouts; improper work methods; improper tools; tool vibration; and job design concerns that include aspects of work flow, line speed, posture and force required, work/rest regiment, and a repetition rate.

7. *Ergonomic Risk Factors* - Conditions of a job, process, or operations that contribute to the risk of developing CTDs are major factors to consider. Examples include repetitiveness of activity force required, and awkwardness of posture. Risk factors are regarded as synergistic effects of ergonomic stressors that must be considered because of their combined effect in developing CTDs. Jobs, operations or workstations that have multiple risk factors will have a higher probability of causing CTDs, depending on the relative degree of severity of each factor.

8. *Posture* - Posture relates to the relative position of the body or body part compared to some standard or reference position (orientation).

9. *Static Work Posture* - Muscle contraction without motion; also, known as isometric work. (Standing is an example of static postural work; gripping or holding is an example of static manual work.)

10. *Work Equipment* - Tools, machines, devices, installations, and other components used in the work environment.

11. *Work Environment* - The work environment is constituted by people and operating equipment acting together in a work process, at the workplace in the work environment

under the conditions imposed by the work task. It must be designed to perform a particular work task safely.

12. *Workplace* - An area allocated to a person in the work environment.

13. *Work Practice* - The sequence of time and space of the interaction of people, work equipment, materials energy, and information within a work environment.

14. *Work Stress (External Load)* - The external conditions and equipments affecting people in the work environment.

Application and Responsibilities

The following section of the program outlines the application requirements of the ergonomic procedure for the following functional responsibilities:

- Department Chief and Other Senior Management/Administration

- Safety Officer and Safety Division (not all departments will have a safety unit)

- Medical Provider

- Ergonomic Committee (or safety division or other dedicated safety unit)

- Purchasing Department or Committee

Senior Management/Administration

Implementing an effective ergonomics program requires top management support and visible involvement at all levels of management, so that all personnel fully understand that management has a serious commitment to the program.

Senior management will be responsible for assigning personnel as ergonomic coordinators and for participating on the ergonomic committee. Management within respective companies or at the station level (line-supervisors) will also be required to ensure that compliance with this procedure is carried out by all employees under their direct supervision. All levels of senior management will be responsible for carrying out the following segments for the ergonomics program:

- Ensure that this policy is implemented and that management's commitment is maintained at all levels of the department.

- Communicate the responsibilities of this policy so that all supervisors and personnel know what is expected.

- Periodically review the program to determine if it is achieving the desired results and is effective.

Safety Officer and Safety Department (or other designated individual or unit)

The safety officer and the safety department is responsible for issuing ergonomic management guidelines to help in achieving ergonomic evaluations; doing regular program reviews; coordinating ergonomic training programs; coordinating they ergonomic committee; overseeing medical management systems; and tracking injury loss trends.

The safety department is responsible for managing the program on a daily basis and ensuring that all segments of the procedure are being followed. Specific responsibilities of Safety may include the following ergonomic program activities and functions:

- Coordinate implementation of the ergonomic program with Senior Management.

- Coordinate with the purchasing department and conduct and arrange for outside ergonomic evaluations of new equipment or workstations identified as having possible ergonomic exposure or other risk factors.

- Coordinate applicable training programs on ergonomics for supervisors and other personnel.

- Coordinate the ergonomics committee and monitor the effectiveness of ergonomic activities.

- Develop and administer an ergonomic monitoring program that will track the status of recommended ergonomically-based department improvements.

- Implement a proactive program to address adverse injury trends, personnel complaints related to ergonomic risk factors, and recognition of an effective workplace ergonomic change instituted.

The ergonomic program should be considered part of the department's overall occupational safety and health program which meets the requirements of NFPA 1500, *Standard on Fire Department Occupational Safety and Health Program.*

Medical Provider

The medical provider (when involved) is responsible for developing and administering a medical management system for the prevention and treatment of cumulative trauma disorders and musculoskeletal injuries related to manual material handling. This program will include the following medical management program elements:

- The medical provider, with other members of the ergonomic committee, will develop a group of jobs that will offer the least ergonomic risk. This group of jobs will be help the medical provider in recommending assignments to light duty or restricted duty jobs as applicable. In addition, they will help personnel rehabilitate and return to their normal work activities when possible.

- Within the safety department, organize a system to track and measure the extent of symptoms of work disorders for each work task. The purpose of this measurement system is to determine which jobs are exhibiting concerns and to aid in measuring the progress of the ergonomic program.

- Maintain a list of qualified health care providers who are have occupational medicine backgrounds and are knowledgeable about fire fighting and emergency medical operations.

- Conduct baseline and follow-on medical examinations or coordinate these examinations with other health care providers.

- Oversee the physical training program for personnel.

- Provide or refer treatment and follow-up procedures appropriate to the injury along with screening and assessments for possible cases of CTD injuries.

- Maintain contact with outside physicians to discuss early return to work options as related to CTD.

- Establish procedures related to avoidance of heat, cold, noise, and visual stress by providing input to the selection process of PPE and other equipment.

Ergonomic Committee

The responsibility of the Ergonomic Committee will be to foster communication on ergonomic concerns and assist with the implementation of the ergonomic program.

Departments are structured uniquely, so it is up to top management to identify the key members of an ergonomics committee. The committee should consist of a carefully selected group of personnel who represent different functions in the department. The committee will be responsible for periodically evaluating the work place and work practices to identify areas that may contribute to musculoskeletal injuries.

Safety officers and safety department personnel must understand and continue to learn techniques for problem identification, training methods, and injury analysis. Managers and supervisors must be trained to interface with line personnel and to recognize potential ergonomic hazards.

If committee members have a special interest in the reduction of CTDs, they will be more committed to address CTD problems in more detail. To be more effective, the ergonomic committee must include representatives from the following groups:

- Department administration: responsible for safety; and providing resources, direction, credibility, and accountability.

- Line Personnel: understand special worker concerns; can communicate with other employees to enlist their acceptance and support of the program.

- Line Supervision: have a special understanding of task demands and the human elements required to get the job done.

- Engineering and supply: familiar with department equipment and tools

- Safety and medical providers: understanding physical and psychological stresses to the body that can result in CTDs.

The direct functions and activities of the committee are as follows:

- Members of the committee will be representatives from each department administration administration, line personnel, line supervision, engineering/ supply, safety/medical providers, and other departments deemed appropriate.

- The safety department will be responsible for coordinating meetings and ensuring that all applicable business segments are represented.

- The committee will appoint a representative to act as the chairperson on a rotating basis with a time schedule established are represented.

- The committee will be responsible for addressing ergonomic issues such as: training, communication, and regulatory compliance,

- The committee will be required to collect information relative to the workplace ergonomic changes instituted and document these changes in a collective data source.

- The committee will periodically review the effectiveness of the program, at least every 3 years.

Training

Committee members must receive the appropriate training to be effective at recognizing and controlling other ergonomic risk factors. If a control program is to be successful, training cannot be limited to the committee members program must be development to include the following:

- Risk factors

- Means of prevention

- Detection of early symptoms

- Importance of reporting symptoms early

- Appropriate work practices.

Personnel with the appropriate training are more likely to be active in controlling CTDs in the workplace.

Job Hazard Analysis

A Job Hazard Analysis (JHA) provides a structured method for identifying potential injuries. The main purpose of a JHA in ergonomics is to identify activities contributing to CTDs and other ergonomic disorders. Data from injury records, medical reports, insurance records, workers' compensation reports, and the OSHA 200 logs (if used) should be analyzed. Indicators such as soreness; pain; strain; or edema of the hand, wrist, elbow, arm, and shoulder may signal possible problems.

Repetitiveness is a known factor. However, the exact number of repetitions in a given time that contribute to CTD is unknown and it will not be the same for each person. Calculation of the repetitiveness of cycle time provides another means for comparing jobs to determine which tasks present greater risks to employees. Jobs with a higher number of repetitions most frequently indicate higher risk.

The next step in analyzing a job is assessing the particular body parts used to perform the task. Review each element and anatomical part for:

- Posture
- Force
- Pressure
- Vibration
- Temperature.

Abnormal posture of any body part in doing the particular task can be a CTD risk. The goal is to maintain body parts in as near a neutral position as possible. Any type of stress, whether created by surface edges rubbing the skin or tool handles pressing into the palm or vibration, has the potential to contribute to CTDs. Vibration can be caused by hand tools or transmitted by motors located under work surfaces. Personnel who work in cold environments, use chemicals that lower the skin temperature, or work with hands exposed to exhaust from tools may also be at risk for CTD. Any job having one or more of these risk factors should be addressed. The following options should be considered for each job element having risk factors:

- Eliminate the task or element if possible

- Redesign the task so that it has satisfactory ergonomic characteristics

- Retrain the worker to complete the task in a way that eliminates the risk

- Provide appropriate PPE or other equipment

- Rotate personnel.

Videotaping personnel during responses or other jobs provides an excellent mechanism for evaluating body position, stress points, and repetitiveness. Slow motion and stop action will give additional advantages for evaluation. Often the actual tasks are performed in rapid motion, making direct visual observation difficult.

Videotapes showing correct work practices and incorrect methods can be useful training tool for employees. Training should be accomplished before employees begin their jobs and periodically thereafter. Practical on-the-job training with other skilled employees is often desirable until new employees become proficient in their jobs. Videotapes of tasks at the workaday can be used to gain skill in JHA and as a training tool. Such tapes will demonstrate both body movements that can lead to injury and alternative movements or positions that may reduce risk.

Once problem risk factors are identified, recommendations for change can be made. Careful planning of job changes is vital. Examples of changes can include:

- Rounding or extending handles to relieve stress on the tissues of the hand.

- Reducing vibration by using dampening materials on the body contact surfaces or replacing the tool with one designed for vibration control.

- Orienting work surfaces so that parts and materials can be positioned without requiring extreme postures of the employees.

- Eliminating unnecessary work elements.

- Structuring response efforts by rotating personnel and by provide rehabilitation on the scene of extended emergency operations.

Whenever any changes are made in the job, watch with careful attention so a new hazard is not created when you are eliminating another. All changes require a period of adjustment and evaluation.

Medical Management

Medical management is used to control and focus on early identification treatment of CTDs and other ergonomic disorders. All personnel must be encouraged to report early signs of CTDs, acute trauma, or other ergonomic disorders. A written medical management system for ergonomics must be developed. This system must be updated periodically and as necessary.

A specific health care provider should be identified which has occupational medicine experience and knowledge of fire fighting and emergency medical operations. Other physicians with specializations within occupational medical should be identified for future referrals.

The medical management system should be responsible for coordination of baseline and follow-on medical examinations. A similar responsibility should include review and coordination of regular fitness training and fitness evaluations.

The department should have a fitness training program which defines appropriate fitness training, supervision, allocation of space, time allocation, and incentives.

Another component of the CTD control program is the management of restricted personnel. Communication between the treating physicians and the workplace task force is vital. They may be able to play a role in specifying any appropriate restrictions. Restricted personnel should be placed in jobs that do not include repetitive motions or abnormal positions. A listing of light duty jobs in each department should be developed by the ergonomic committee with department supervisors. This listing should be shared with the treating physicians. In addition, a periodic review of work restrictions is necessary to manage personnel properly. Educating all personnel involved with restricted employees will help ensure success.

Evaluations of Program Effectiveness

To be effective, an ergonomic program must be constantly evaluated and updated as necessary. One form of evaluation is to compare the CTD incidence rate after implementing the control program with the rate for the year before the program. A more sophisticated evaluation technique is to conduct a cost/benefit analysis, particularly at some point after the program has been implemented. Early attention is being given to the problem, which may result in increased identification of CTDs. Constant monitoring is necessary to ensure that personnel with CTDs are not removed from their jobs and placed into other jobs that have the same risk factors.

Tool Design

Tools used during response operations and at the station must be well-suited for the personnel and required tasks. For hand tools, the design should allow for:

- neutral wrist posture,

- reduce vibration transmissions to the hand and upper extremities,

- provide the proposed mechanical advantages, and

- possess other characteristics to intended reduce individual injury potential.

Factors, such as the amount of bend in a handle and reach (extension) are highly task specific. Optimal tool design characteristics can only be determined because of data developed at the workplace or in a laboratory evaluation that simulates task requirements.

This quantitative assessment allows for the proper identification of specific tools that produce cumulative wrist injuries and provides essential data for developing alternative work methods and tool designs to reduce injuries.

The object is to make the tool fit the fire fighter. These include weight, shape, vibration, noise, grips, switches, application pressure, and operator posture. Even more complex is the process of determining what is most ergonomically correct.

The design process begins with input from tool users and is not completed until a prototype of the new tool is put into the fire fighter's hands.

Use of PPE

PPE should be selected on the basis of human factors considerations as well as protective features. When possible PPE should be:

- as light as possible,

- not restrict fire fighter (or EMT) movement,

- be appropriate sized (and available in sufficient sizing for correct fit of department personnel, and

- function as claimed by the manufacturer.

Since 'breathable' protective clothing is known to provide lower stress on individuals during light and moderate work loads, protective clothing with 'breathable' moisture barriers should be selected.

Improper footwear and lack of ankle support are major causes of ankle and lower extremity injuries. Footwear which provides good fit to the individual, especially ankle support, and having wide soles should be selected over poorly designed, inadequately designed footwear which does not provide normal fit and ankle support.

APPENDIX D

SAMPLE FITNESS EVALUATION PROTOCOL

(Provided by the Phoenix Fire Department
as part of the Health Center Manual)

TABLE OF CONTENTS
FOR
FITNESS EVALUATION

FITNESS EVALUATION

Introduction

In combination with the physical examination each member of the Phoenix Fire Department is offered the opportunity to participate in the fitness evaluation. Evaluations will be conducted in six areas that are vital to total fitness.

1. Flexibility
2. Muscular Power
3. Muscular Endurance
4. Muscular Strength
5. Body Composition
6. Aerobic Capacity

The evaluation is designed to help fire-fighters make informed decisions about personal health issues.

Evaluation and Preparation Methods

Flexibility

Definition - Flexibility is, "The ability of a joint to go through a full range of motion." It is historically the most neglected area of fitness. Ironically, it is recognized as one of the most important factors contributing to both enhanced performance and injury prevention.

Evaluation Methods - The sit and reach, the shoulder rotation, and the trunk rotation tests will be used to evaluate flexibility.

Preparation - Make it part of the every day routine to incorporate in some basic stretching exercises.

Muscular Power

Definition - Muscular power is strength in relationship to time. It is also known as "explosive" strength. The very nature of fire suppression requires at times the use of muscular power.

Evaluation Method - Sit-ups, pull-ups, dips and push-ups will be used to measure endurance. They will all be maximum efforts, with no time limit. Each must be performed in a strict, continuous manner. Fire-fighters are offered the opportunity to perform a 80 pound bench press in place of push-ups and a 100 pound anterior lat pulldown in place of pull-ups.

Preparation - Weight training with emphasis on high repetition can be beneficial. Like the vertical jump, the best method to train for muscular endurance is to practice the actual exercises used in the evaluation.

Muscular Strength

Definition - Muscular strength is similar to power, except there is no time factor.

Evaluation Method - The bench press will be used to evaluate the upper body. The 45 degree leg press will be used to evaluate the lower body. A jamar hand dynamometer will be used to measure grip strength.

Preparation - A progressive weight training program built around the bench and leg press will build sufficient strength. Specific activities for enhanced grip strength must also be included.

Body Composition

Definition - The crucial measurement here is the percent body fat. The amount of body fat carried around has been shown to increase a persons risk of cardia-vascular disease and muscular-skeletal injuries. Physical performance is also compromised.

Evaluation Method - Skin fold measurements will be taken from the triceps, chest, abdomen, thigh, and back, using a caliper. Measurements obtained are manipulated in a formula to obtain the percent body fat. While this method is subject to error, it is economically and practically speaking the best method available at this time. It should be noted that because there is a degree of error in this method it may be more appropriate not to attempt any evaluation, but instead use photographs to visually show changes that occur from year to year.

Preparation - Activities should be pursued that are considered aerobic in nature. See aerobic power for more information.

Aerobic Power

Definition - The condition of the cardio-vascular system and the pulmonary system is measured by aerobic power. It measures how effective the body is at delivering oxygen to the working muscle.

Evaluation Method - The motorized treadmill using the Balke protocol is the current method used to determine aerobic power.

Preparation - Aerobic exercise is exercise that elevates your heat to a specified level (target hear rate) and maintains it there for a minimum of twenty minutes. Jogging, swimming, biking, and rowing are just a few examples. Aerobic exercise should be performed three times weekly.

Summary

The fitness evaluation is only one part of the fitness program. It will provide useful information for the fire-fighter. It is information that will allow for informed decisions to be made in respect to the quality of life one experiences.

Sit and Reach

Objective: To measure the flexibility of the lower back and the muscles located at the back of the thigh (hamstrings).

Directions:

1. The participant should be instructed to sit on the floor with his/her back against the wall and legs fully extended and together in front.

2. The participant should then be instructed to sit up straight.

3. The sit and reach measuring device will then be placed into position at the feet of the participant.

4. Instruct participant to reach forward with full extended arms, right hand on top of left hand, but keeping shoulder blades firmly against the wall.

5. The sliding measuring tape should be adjusted in such a manner that the starting point of the tape is at the fingertips of the outstretched arms of the participant. This procedure will compensate for differences in arm length, allowing for a more accurate assessment of flexibility.

6. The participant then is instructed to slowly strength forward, bending at the waist, keeping legs fully extended.

7. The participant must hold the strength for a period of three seconds to be acceptable.

8. The best of three attempts will be used,

Vertical Jump

Objective: To measure the power of the lower extremities.

Directions:

1. The participant stands with feet shoulder length apart and with one shoulder against the wall.

2. A standard marker is held in the hand closest to the wall.

3. The participant will reach up as high as possible while keeping feet flat on the ground and make a mark on the wall.

4. This mark will be the reference point from which the height of the jump will be determined.

5. The participant will then be instructed to jump as high as possible and make another mark on the wall.

6. The participant should be allowed three trials.

7. The best effort should be used to determine score.

8. The distance between the reference mark and the best trial mark will be the participant's score. Measurement will be in inches.

Pull-ups

Objective: To measure upper body muscular endurance.

Directions:

1. Participant should be instructed to take an approximately shoulder width, palms away grip.

2. Participant should not be touching ground with feet. If necessary, participant may bend legs at the knees.

3. Participant must raise his/her body to a sufficient height to allow him to touch the bar with his chin.

4. The participant should lower his body down to a point where full extension of the elbow is achieved.

5. The pull-ups should be performed in a controlled manner. No kicking will be allowed.

6. This will be a maximum effort.

    There is no time limit.

100 Pound Anterior Lat Pull-Down

Objective: To measure endurance of upper body.

Directions:

1. The participant will grab a standard lat bar with a palms away, shoulder width grip.

2. The weight lifted should total 100 pounds.

3. The participant sitting in an erect position will begin the movement by pulling the lat pull-down to a point where it comes in contact with the clavicle at the mandibular notch. The weight will then be returned to the starting point. This sequence of movements will be repeated as many times as possible.

4. The participant must perform the movement in a strict fashion, maintaining control over the weight at all times.

Push-Ups

Objective: To measure endurance of upper body.

Directions:

1. The participant will start from the up position.

2. Hands should be placed on floor at a comfortable width.

3. The body should be held in a straight line with arms fully extended.

4. The participant should lower his/her body until the chest touches the hand of the evaluator.

5. The participant will then "push-up" and return to the starting position where arms are fully extended.

6. This movement will be repeated as many times as possible without resting.

7. There is no time limit.

80 Pounds Bench Press

Objective: To measure endurance of upper body.

Directions:

1. The participant will lay down on bench with feet flat on the ground.

2. A Olympic bar and weights totaling 80 pounds will be used.

3. The participant will grab the bar with a comfortable width grip and begin by lowering weight down to the chest. The weight will then be returned to the starting position by extending arms upward.

4. The lowering and raising of the weights must be performed at a cadence of one second intervals. For example, the bar will be lowered and touch the chest on the first cadence. The weight must then be returned to the starting position with arms fully extended by the second cadence. This sequence is repeated as many times as possible until the participant is unable to keep up with the cadence.

5. Bar must not be bounced off chest.

6. Arms must be fully extended.

7. Arching of lower body not permitted.

Sit-Ups

Objective: To measure the endurance of the abdominal muscles.

Directions:

1. Sit-up board should be positioned on the floor.

2. Participant should be instructed to position self on sit-up board.

3. Both feet should be secured under forward pad.

4. The second pad should be individually adjusted to fit behind the knees of the participant.

5. The participant should be instructed to either interlace the fingers behind the head or cross arms across chest. The shoulder blades must be in contact with the board.

6. Participant should roll up bringing his/her chest as close to their knees as possible. Participant then will lower his body back to the starting position with shoulder blades making contact with the board.

7. Repeat the above described movement as many times as possible.

8. The participant will not be allowed to stop or pause in the top or bottom position,

9. The total number of repetitions performed correctly will be recorded.

10. There is no time limit.

Bar Dips

Objective: To evaluate upper body muscular endurance.

Directions:

1. The participant will start in the top position with arms fully extended.

2. The participant will then lower his/her body to a point where the upper arm is parallel to the floor.

3. The participant will repeat this movement as many times as possible.

4. The participant will be allowed to stop in the top position only.

5. The participant must perform dips in a strict fashion. Kicking with the legs will not be allowed.

Hand Grip Strength Test

Objective: To measure hand grip strength

Directions:

1. This test is performed with the jamar hand grip dynamometer.

2. Adjust dynamometer to fit comfortably in participant's hand.

3. The participant may stand or sit.

4. The shoulder be adducted and neutrally rotated.

5. The elbow should be flexed to 90 degrees.

6. The forearm and wrist should be in the neutral position.

7. The participant should then be instructed to squeeze as hard as possible.

8. Three attempts should be performed and then averaged to give a final score.

Upper Body Strength Test

Objective: To measure upper body strength.

Directions:

1. The bench press will be the exercise used to assess upper body strength.

2. The bench press will be performed in accordance with accepted power-lifting rules and regulations.

3. The participants should be allowed to warm-up in any manner he or she wants.

4. The participants should have both feet flat on the floor during the lift.

5. The buttocks, shoulder blades, and back of head must remain in contact with bench at all times.

6. The weight must be controlled on the way down.

7. There must be a momentary pause once the bar contacts the chest, Bouncing the bar will not be accepted.

8. The weight must then be pushed up in a controlled manner.

9. The bench press will be a maximum effort.

10. The repetitions achieved should be within the range of one to five.

Lower Body Strength Test

Objective: To measure the strength of the lower body.

Directions:

1. The 45 degree leg press will be the exercise used to assess lower body strength.

2. The participant should be instructed to place feet on push plate approximately shoulder width apart.

3. The participant will then be instructed to lower carriage (prior to loading with weights) to a point where the knee joint is bent at a 90 degree angle or more. The evaluator will mark this point with a marker. This mark will then serve as a guide to ensure proper execution of the lift.

4. Participant should be allowed to warm-up in the manner he/she feel most comfortable.

5. The lift should be performed in a controlled manner. No bouncing.

6. The weight then should be lowered and returned to the starting position with the legs fully extended.

7. The leg press will be a maximum effort.

8. The repetitions achieved should be within the range of one to five.

Body Fat Analysis (Skin Fold Technique)

This technique measures subcutaneous fat with calipers at selected sites on the body. The values obtained are reported in millimeters and manipulated in a formula to arrive at the percent body fat. This technique is relatively inexpensive to administer, but historically has been plagued with human error making the data obtained unreliable. Training classes and periodic refresher courses can alleviate this concern somewhat, but not completely.

There exists a host of skin fold techniques. The one illustrated below is the technique currently being used today by the Phoenix Fire Department.

1. Instruct fire-fighter to disrobe down to his physical training shorts. Female fire-fighters will obviously be evaluated by a female staff member.

2. Take average of measurements on both right and left sides. The sites are;
   - Scapula inferior border
   - Pectoralis midway between should and opposite nipple
   - Triceps midway between elbow and shoulder
   - Abdomen one inch from navel laterally
   - Thigh midway between knee and hip, weight should be on opposite leg.
   - Supra-ilium, mid-axillary line.

3. Add the measurements of the body fold in each area:
   - X2=SUM of triceps, suprailium, and thigh skinfolds.
   - X3=SUM of chest, abdomen, and thigh skinfolds.
   - X4=SUM of chest, triceps, and scapular skinfolds.
   - XS=SUM of triceps, suprailium, and abdomen skinfolds.
   - X6=Age in years

4. For males, use the average of the two following formulas
   $$B.D. = 1.1093800 - 0.0008267 (X3) + 0.0000016 (X3)^2 - 0.0002574 (X6)$$
   $$B.D. = 1.1125025 - 0.0013125 (X4) + 0.0000055 (X4)^2 - 0.002440 (X6)$$

5. For females, use the average of the two following formulas
   $$B.D. = 1.0994821 - 0.0009929 (X2) + 0.0000023 (X2)^2 - 0.0001392 (X6)$$
   $$B.D. = 1.0902369 - 0.0009379 (X5) + 0.0000026 (X5)^2 - 0.0001087 (X6)$$

6. Calculate bodyfat from either of the two following equations
   Brozek et al. = [4.570/Body Density) - 4.1421 x 100
   SIRI= [(4.950/Body Density) - 4.5001 x 100

Treadmill Protocol

Objective: To measure individual cardiovascular endurance

Directions:

1. The treadmill protocol used currently by Phoenix Fire Department is the Balke-C. Other treadmill protocols may be acceptable if the same measurements are made.

2. The test is administered by trained individuals, who are under direct supervision by the health center physician on duty.

3. The fire-fighter is monitored by a standard 12 lead ECG.

4. The Balke protocol calls for a one percent change in grade for each of the first twenty-five minutes. The speed is held constant at 3.3 mph during this initial phase.

5. The second phase is characterized by a constant percent grade of 25 percent, and a increase in speed of two-tenths of a mph for every minutes the fire-fighter remains on the treadmill.

6. This protocol is a maximal effort.

Air Dyne Protocol

Objective: To measure aerobic power.

Equipment:

1. Air dyne bike
2. Stop watch

Procedure:

1. First obtain body weight of participant

2. Adjust seat height.

3. Refer to Table I to determine workload for each stage.

4. Begin exercising.

   a. Stages 1, 2, 3, are each 4 minutes in duration. Each stage there after is only one minute in duration.

   b. Remember this is a maximum effort.

5. Record elapsed time. Round off to nearest 20 second interval. For example, a time of 10 minutes, 16 seconds will be rounded off to 10 minutes, 20 seconds.

6. Refer to Tables 3a, 3b, and 3c, to obtain final work load performed.

7. To determine participants max $V(0_2)$ use the following formula;

   a. Y = Final work load performed

   b. $V(0_2) = 3.5 + [(Y \times 1320)/\text{body weight in pounds}]$

Table 3a. Work Load Performed during Air Dyne Protocol

Elapsed Time

Bod. Wt.	4:00	4:20	4:40	5:00	5:20	5:40	6:00	6:20	6:40	7:00	7:20	7:40	8:00	8:20	8:40	9:00	9:20
120	1.2	1.3	1.4	1.5	1.6	1.7	1.8	1.9	2.0	2.1	2.2	2.3	2.4	2.5	2.6	2.7	2.7
125	1.3	1.4	1.5	1.6	1.7	1.8	1.9	2.0	2.1	2.2	2.3	2.4	2.5	2.6	2.7	2.8	2.9
130	1.3	1.4	1.5	1.6	1.7	1.8	1.9	2.0	2.2	2.2	2.4	2.5	2.6	2.7	2.8	2.9	3.0
135	1.4	1.5	1.6	1.7	1.8	1.9	2.0	2.1	2.2	2.3	2.5	2.6	2.7	2.8	2.9	3.0	3.1
140	1.4	1.5	1.6	1.7	1.9	2.0	2.1	2.2	2.3	2.4	2.5	2.6	2.7	2.8	3.0	3.1	3.2
145	1.4	1.6	1.7	1.8	1.9	2.0	2.2	2.2	2.4	2.5	2.6	2.8	2.9	3.0	3.2	3.3	3.4
150	1.5	1.6	1.7	1.8	1.9	2.0	2.2	2.3	2.4	2.6	2.7	2.8	3.0	3.1	3.2	2.4	2.5
155	1.6	1.7	1.8	1.9	2.0	2.1	2.3	2.4	2.5	2.7	2.8	2.9	3.1	3.2	3.3	3.5	3.6
160	1.6	1.7	1.8	2.0	2.1	2.3	2.4	2.5	2.6	2.8	2.9	3.0	3.2	3.3	3.4	3.6	3.7
165	1.7	1.8	1.9	2.1	2.2	2.3	2.5	2.6	2.7	2.9	3.0	3.1	3.3	3.4	3.5	3.7	3.8
170	1.7	1.8	2.0	2.1	2.3	2.4	2.5	2.6	2.8	2.9	3.0	3.2	3.3	3.4	3.6	3.8	3.9
175	1.8	1.9	2.0	2.2	2.3	2.4	2.6	2.7	2.8	3.0	3.1	3.3	3.4	3.5	3.7	3.9	4.0
180	1.8	1.9	2.1	2.2	2.4	2.5	2.7	2.8	2.9	3.1	3.2	3.3	3.5	3.6	3.8	4.0	4.1
185	1.8	2.0	2.1	2.3	2.4	2.5	2.7	2.8	3.0	3.2	3.3	3.4	3.6	3.7	3.9	4.1	4.2
190	1.9	2.1	2.2	2.3	2.4	2.6	2.8	2.9	3.1	3.3	3.4	3.6	3.8	3.9	4.0	4.1	4.3
195	2.0	2.1	2.2	2.4	2.5	2.7	2.9	3.0	3.2	3.4	3.5	3.6	3.8	3.9	4.1	4.3	4.4
200	2.0	2.2	2.3	2.5	2.6	2.7	2.9	3.1	3.2	3.4	3.5	3.7	3.9	4.0	4.2	4.4	4.5
205	2.1	2.3	2.4	2.6	2.7	2.8	3.0	3.1	3.3	3.5	3.6	3.8	4.0	4.1	4.3	4.5	4.6
210	2.1	2.3	2.4	2.6	2.7	2.9	3.1	3.2	3.4	3.6	3.7	3.9	4.1	4.2	4.4	4.6	4.7

Table 3a. Work Load Performed during Air Dyne Protocol (continued)

Elapsed Time

Bod. Wt.	4:00	4:20	4:40	5:00	5:20	5:40	6:00	6:20	6:40	7:00	7:20	7:40	8:00	8:20	8:40	9:00	9:20
215	2.2	2.4	2.5	2.7	2.8	3.0	3.2	3.3	3.5	3.7	3.8	4.0	4.2	4.3	4.5	4.7	4.9
220	2.2	2.4	2.6	2.7	2.8	3.0	3.2	3.3	3.5	3.7	3.9	4.1	4.3	4.4	4.6	4.8	5.0
225	2.3	2.5	2.6	2.8	2.9	3.1	3.3	3.4	3.6	3.8	4.0	4.2	4.4	4.5	4.7	4.9	5.1
230	2.3	2.5	2.7	2.8	3.0	3.2	3.4	3.5	3.7	3.9	4.1	4.3	4.5	5.6	4.8	5.0	5.2
235	2.4	2.6	2.7	2.9	3.1	3.3	3.5	3.6	3.8	4.0	4.2	4.4	4.6	4.7	4.9	5.1	5.3
240	2.4	2.6	2.8	3.0	3.2	3.4	3.5	3.7	3.9	4.1	4.3	4.5	4.7	4.8	5.0	5.2	5.4
245	2.5	2.7	2.8	3.0	3.2	3.4	3.6	3.8	4.0	4.2	4.4	4.6	4.8	5.0	5.2	5.4	5.6
250	2.5	2.7	2.9	3.1	3.3	3.5	3.7	3.9	4.1	4.3	4.5	5.7	4.9	5.1	5.3	5.5	5.7
255	2.6	2.8	3.0	3.2	3.4	3.6	3.8	4.0	4.2	4.4	4.6	4.8	5.0	5.2	5.4	5.6	5.8
260	2.6	2.8	3.0	3.2	3.4	3.6	3.8	4.0	4.2	4.4	4.6	4.8	5.1	5.3	5.5	5.7	5.9
265	2.7	2.9	3.1	3.3	3.5	3.7	3.9	4.1	4.3	4.5	4.7	4.9	5.2	5.4	5.6	5.8	6.0
270	2.7	2.9	3.1	3.3	3.5	3.7	4.0	4.2	4.4	4.6	4.8	5.0	5.3	5.5	5.7	5.9	6.1
275	2.8	3.0	3.2	3.4	3.6	3.8	4.1	4.3	4.5	4.7	4.9	5.1	5.4	5.6	5.8	6.0	6.2
280	2.8	3.1	3.3	3.5	3.7	3.9	4.1	4.3	4.4	4.8	5.0	5.2	5.5	5.7	5.9	6.1	6.3
285	2.9	3.2	3.3	3.5	3.7	3.9	4.2	4.4	4.6	4.9	5.1	5.3	5.6	5.8	6.0	6.3	6.5
290	2.9	3.2	3.4	3.6	3.8	4.0	4.3	4.5	4.7	5.0	5.2	5.4	4.7	5.9	6.1	6.4	6.6
295	3.0	3.2	3.4	3.7	3.9	4.1	4.4	4.6	4.8	5.1	5.3	5.6	5.8	6.0	6.2	6.5	6.7

Table 3b. Work Load Performed during Air Dyne Protocol

Bod. Wt.	9:40	10:00	10:20	10:40	11:00	11:20	11:40	12:00	12:15	12:30	12:45	13:00	13:15	13:30	13:45	14:00	14:15
											Elapsed Time						
120	2.8	2.9	3.0	3.1	3.2	3.3	3.4	3.5	3.6	3.7	3.9	4.0	4.1	4.3	4.4	4.5	4.6
125	3.0	3.0	3.1	3.2	3.3	3.4	3.5	3.6	3.7	3.8	3.9	4.0	4.1	4.3	4.4	4.5	4.6
130	3.1	3.2	3.3	3.4	3.5	3.6	3.7	3.8	3.9	4.1	4.3	4.5	4.6	4.8	4.9	5.0	5.1
135	3.2	3.3	3.4	3.5	3.6	3.7	3.8	3.9	4.0	4.1	4.3	4.5	4.6	4.8	4.9	5.0	5.1
140	3.3	3.4	3.5	3.6	3.7	3.8	3.9	4.0	4.1	4.2	4.4	4.5	4.6	4.8	4.9	5.0	5.1
145	3.4	3.5	3.6	3.7	3.9	4.0	4.1	4.2	4.2	4.3	4.4	4.5	4.6	4.8	4.9	5.0	5.1
150	3.6	3.7	3.8	3.9	4.1	4.2	4.3	4.4	4.5	4.6	4.8	5.0	5.1	5.3	5.4	5.5	5.6
155	3.7	3.8	3.9	4.0	4.2	4.3	4.4	4.6	4.7	4.8	4.9	5.0	5.1	5.3	5.4	5.5	5.6
160	3.8	3.9	4.0	4.1	4.3	4.4	4.5	4.7	4.8	4.8	4.9	5.0	5.1	5.3	5.4	5.5	5.6
165	3.9	4.1	4.2	4.3	4.5	4.6	4.7	4.8	4.9	5.1	5.3	5.5	5.6	5.8	5.9	6.0	6.1
170	4.0	4.2	4.3	4.4	4.6	4.7	4.8	5.0	5.1	5.2	5.4	5.5	5.6	5.8	5.9	6.0	6.1
175	4.1	4.3	4.4	4.5	4.7	4.8	4.9	5.1	5.2	5.3	5.4	5.5	5.6	5.8	5.9	6.0	6.1
180	4.2	4.4	4.5	4.6	4.8	4.9	5.0	5.2	5.3	5.3	5.4	5.5	5.6	5.8	5.9	6.0	6.1
185	4.3	4.5	4.6	4.8	5.0	5.1	5.2	5.4	5.5	5.6	6.8	6.0	6.1	6.3	6.4	6.5	6.6
190	4.5	4.7	4.8	4.9	5.1	5.2	5.4	5.6	6.7	5.8	5.9	6.0	6.1	6.3	6.4	6.5	6.6
195	5.6	4.8	4.9	5.0	5.2	5.3	5.5	5.7	5.8	5.8	5/9	6.0	6.1	6.3	6.4	6.5	6.6
200	4.7	4.9	5.0	5.1	5.3	5.4	5.6	5.8	5.9	6.1	6.3	6.5	6.6	6.8	6.9	7.0	7.1

Table 3b. Work Load Performed during Air Dyne Protocol (Continued)

Bod. Wt.	9:40	10:00	10:20	10:40	11:00	11:20	11:40	12:00	12:15	12:30	12:45	13:00	13:15	13:30	13:45	14:00	14:15
205	4.8	5.0	5.1	5.3	5.5	5.6	5.8	6.0	6.1	6.2	6.4	6.5	6.6	6.8	6.9	7.0	7.1
210	4.9	5.1	5.2	5.4	5.6	5.7	5.9	6.1	6.2	6.3	6.4	6.5	6.6	6.8	6.9	7.0	7.1
215	5.1	5.3	5.4	5.6	5.8	5.9	6.1	6.3	6.4	6.6	6.8	7.0	7.1	7.2	7.4	7.5	7.6
220	5.2	5.4	5.5	5.7	5.9	6.0	6.2	6.4	6.5	6.6	6.8	7.0	7.1	7.3	7.4	7.5	7.6
225	5.3	5.5	5.7	5.9	6.1	6.2	6.4	6.6	6.7	6.8	6.9	7.0	7.2	7.3	7.4	7.5	7.6
230	5.4	5.6	5.8	6.0	6.2	6.4	6.6	6.8	6.9	7.1	7.3	7.5	7.6	7.8	7.9	8.0	8.1
235	5.5	5.7	5.9	6.1	6.3	6.5	6.7	6.9	7.0	7.1	73	7.5	7.6	7.8	7.9	8.0	8.1
240	5.6	5.8	6.0	6.2	6.5	6.6	6.8	7.0	7.1	7.2	7.4	7.5	7.6	7.8	7.9	8.0	8.1
245	5.8	6.0	6.2	6.4	6.6	6.8	6.9	7.1	7.2	7.3	7.4	7.5	7.6	7.8	7.9	8.0	8.1
250	5.9	6.1	6.3	6.5	6.7	6.9	7.1	7.3	7.4	7.6	7.8	8.0	8.1	8.3	8.4	8.5	8.6
255	6.0	6.2	6.4	6.6	6.8	7.0	7.2	7.4	7.5	7.6	7.8	8.0	8.1	8.3	8.4	8.5	8.6
260	6.1	6.3	6.5	6.7	7.0	7.1	7.4	7.6	7.7	7.8	7.9	8.0	8.1	8.3	8.4	8.5	8.6
265	6.2	6.4	6.7	6.9	7.1	7.3	7.5	7.7	7.7	7.8	7.9	8.0	8.1	8.3	8.4	8.5	8.6
270	6.3	6.6	6.8	7.0	7.3	7.5	7.7	7.9	8.0	8.1	8.3	8.5	8.6	8.8	8.9	9.0	9.1
275	6.4	6.7	6.9	7.1	7.4	4.6	7.8	8.0	8.1	8.2	8.3	8.5	8.6	8.8	8.9	9.0	9.1
280	6.5	6.8	7.0	7.2	7.5	7.7	7.9	8.0	8.2	8.3	8.4	8.5	8.6	8.8	8.9	9.0	9.1
285	6.7	7.0	7.2	7.4	7.7	7.9	8.1	8.3	8.4	8.6	8.8	9.0	9.1	9.3	9.4	9.5	9.6
290	6.8	7.1	7.3	7.5	7.8	8.0	8.2	8.5	8.6	6.7	8.8	9.0	9.1	9.3	9.4	9.5	9.6

Elapsed Time

Table 3c. Work Load Performed during Air Dyne Protocol

Bod. Wt.	Elapsed Time																
	14:30	14:45	15:00	15:15	15:30	15:45	16:00	16:15	16:30	16:45	17:00	17:15	17:30	17:45	18:00	18:15	18:30
120	4.8	4.9	5.0	5.1	5.3	5.4	5.5	5.6	5.8	5.9	6.0	6.1	6.4	6.4	6.5	6.6	6.8
125	4.8	4.9	5.0	5.1	5.3	5.4	5.5	5.6	5.8	5.9	6.0	6.1	6.4	6.4	6.5	6.6	6.8
130	5.3	5.4	5.5	5.6	5.8	5.9	6.0	6.1	6.3	6.4	6.5	6.6	6.8	6.9	7.0	7.2	7.3
135	5.3	5.4	5.5	5.6	5.8	5.9	6.0	6.1	6.3	6.4	6.5	6.6	6.8	6.9	7.0	7.2	7.3
140	5.3	5.4	5.5	5.6	5.8	5.9	6.0	6.1	6.3	6.4	6.5	6.6	6.8	6.9	7.0	7.2	7.3
145	5.3	5.4	5.5	5.6	5.8	5.9	6.0	6.1	6.3	6.4	6.5	6.6	6.8	6.9	7.0	7.2	7.3
150	5.8	5.9	6.0	6.1	6.3	6.4	6.5	6.6	6.8	6.9	7.0	7.1	7.3	7.4	7.5	7.6	7.8
155	5.8	5.9	6.0	6.1	6.3	6.4	6.5	6.6	6.8	6.9	7.0	7.1	7.3	7.4	7.5	7.6	7.8
160	5.8	5.9	6.0	6.1	6.3	6.4	6.5	6.6	6.8	6.9	7.0	7.1	7.3	7.4	7.5	7.6	7.8
165	6.3	6.4	6.5	6.6	6.8	6.9	7.0	7.1	7.3	7.4	7.5	7.6	7.8	7.9	8.0	8.1	8.3
170	6.3	6.4	6.5	6.6	6.8	6.9	7.0	7.1	7.3	7.4	7.5	7.6	7.8	7.9	8.0	8.1	8.3
175	6.3	6.4	6.5	6.6	6.8	6.9	7.0	7.1	7.3	7.4	7.5	7.6	7.8	7.9	8.0	8.1	8.3
180	6.3	6.4	6.5	6.6	6.8	6.9	7.0	7.1	7.3	7.4	7.5	7.6	7.8	7.9	8.0	8.1	8.3
185	6.8	6.9	7.0	7.1	7.3	7.4	7.5	7.6	7.8	7.9	8.0	8.1	8.3	8.4	8.5	8.6	8.8
190	6.8	6.9	7.0	7.1	7.3	7.4	7.5	7.6	7.8	7.9	8.0	8.1	8.3	8.4	8.5	8.6	8.8
195	6.8	6.9	7.0	7.1	7.3	7.4	7.5	7.6	7.8	7.9	8.0	8.1	8.3	8.4	8.5	8.6	8.8
200	7.3	7.4	7.5	7.6	7.8	7.9	8.0	8.1	8.3	8.4	8.5	8.6	8.8	8.9	9.0	9.2	9.3
205	7.3	7.4	7.5	7.6	7.8	7.9	8.0	8.1	8.3	8.4	8.5	8.6	8.8	8.9	9.0	9.2	9.3
210	7.3	7.4	7.5	7.6	7.8	7.9	8.0	8.1	8.3	8.4	8.5	8.6	8.8	8.9	9.0	9.2	9.3

Table 3c. Work Load Performed during Air Dyne Protocol (Continued)

Bod. Wt.	\	\	\	\	\	\	\	Elapsed Time	\	\	\	\	\	\	\	\	\
	14:30	14:45	15:00	15:15	15:30	15:45	16:00	16:15	16:30	16:45	17:00	17:15	17:30	17:45	18:00	18:15	18:30
215	7.8	7.9	8.0	8.1	8.3	8.4	8.5	8.6	8.8	8.9	9.0	9.1	9.3	9.4	9.5	9.6	9.8
220	7.8	7.9	8.0	8.1	8.3	8.4	8.5	8.6	8.8	8.9	9.0	9.1	9.3	9.4	9.5	9.6	9.8
225	7.8	7.9	8.0	8.1	8.3	8.4	8.5	8.6	8.8	8.9	9.0	9.1	9.3	9.4	9.5	9.6	9.8
230	8.3	8.4	8.5	8.6	8.8	8.9	9.0	9.1	9.3	9.4	9.5	9.6	9.8	9.9	10.0		
235	8.3	8.4	8.5	8.6	8.8	8.9	9.0	9.1	9.3	9.4	9.5	9.6	9.8	9.9	10.0		
240	8.3	8.4	8.5	8.6	8.8	8.9	9.0	9.1	9.3	9.4	9.5	9.6	9.8	9.9	10.0		
245	8.3	8.4	8.5	8.6	8.8	8.9	9.0	9.1	9.3	9.4	9.5	9.6	9.8	9.9	10.0		
250	8.8	8.9	9.0	9.1	9.3	9.4	9.5	9.6	9.8	9.9	10.0						
255	8.8	8.9	9.0	9.1	9.3	9.4	9.5	9.6	9.8	9.9	10.0						
260	8.8	8.9	9.0	9.1	9.3	9.4	9.5	9.6	9.8	9.9	10.0						
265	8.8	8.9	9.0	9.1	9.3	9.4	9.5	9.6	9.8	9.9	10.0						
270	9.3	9.4	9.5	9.6	9.8	9.9	10.0										
275	9.3	9.4	9.5	9.6	9.8	9.9	10.0										
280	9.3	9.4	9.5	9.6	9.8	9.9	10.0										
285	9.8	9.9	10.0														
290	9.8	9.9	10.0														
295	9.8	9.9	10.0														

D-25

Standards

Minimum Company Standards

There have been made attempts to develop a correlation between certain physical activities and fire-fighting tasks. It is the Phoenix Fire Department's position that more research needs to be conducted in this area before standards can be developed. At the time of this writing actual job skills performed under simulated fire conditions still serve as minimum company standards (MCS).

MCS is conducted annually at the beginning of the calendar year. SCBA evaluations, search and rescue drill, ancillary equipment review, ladders, and protective equipment inspection are performed at this time. The center of attention is the fire drill. The fire drill will simulate fire-ground conditions as close as possible. The crews will be evaluated on all phases of fire ground activity. Performing tasks safely and within the allotted time are common bench marks that. must be met for a passing grade. If a company fails MCS, the errors committed will be discussed and the company will be re-evaluated at a later time. If the errors were a result of poor fitness levels, the fitness coordinator will work with the company officer in designing a program to correct the deficiencies.

APPENDIX E

WORKERS' COMPENSATION LAWS BY STATE

WORKERS' COMPENSATION LAWS BY STATE

In most states, volunteer fire fighters are deemed public employees for the purposes of workers' compensation. Louisiana, Massachusetts, Mississippi, Montana, New Mexico, Oregon, Rhode Island, South Carolina, and Tennessee do not have mandatory coverage for volunteer fire fighters; however, counties or municipalities may be able to elect coverage for volunteers. For volunteers not covered under Workers' Compensation, other legal remedies may be available.

For those covered under Workers' Compensation, methods for calculating distributions vary greatly by state. Some states consider the fire fighter's full-time job. Other states determine compensation based on the paid fire fighter position comparable to the volunteer fire fighter's position. Still others pay an amount based on the state's average wage or the maximum or minimum allowable under the workers' compensation statutes. (One should note that the methods for calculating distributions are not necessarily the same as those used to calculate insurance premiums.)

Below is a table summarizing the compensation method used by each state, as of December 1995. This information, of course, is subject to change; in addition, for an injury occurring prior to this time, the different rules may apply. The workers' compensation administration office should be contacted for specifics concerning coverage for a particular injury date. In most states, a legal advisor is available in the Workers' Compensation office to answer questions.

STATE/JURISDICTION	PHONE NUMBER	COMMENTS
ALABAMA	910/242-2868	benefits for volunteers based on 66 2/3 % of regular job or minimum wage, whichever is greater
ALASKA	907/465-2790	benefits for volunteers based on minimum gross weekly earnings of paid fire fighters
ARIZONA	602/542-4411	benefits for volunteers based on beginning wage of same rank or grade full-time fire fighter
ARKANSAS	800/622-4472	volunteers must file with county clerk to show that they've had minimum training and are member of nonmunicipal fire department; payments based on average county wage
CALIFORNIA	415/703-3731	benefits based on maximum allowable under Workers' Compensation statutes

WORKERS' COMPENSATION LAWS BY STATE (Continued)

STATE/JURISDICTION	PHONE NUMBER	COMMENTS
COLORADO	303/575-8700	benefits based on maximum allowable under Workers' Compensation statutes
CONNECTICUT	203/493-1500	benefits for volunteers based on the CT average weekly wage for production workers in the manufacturing industry as determined by the labor commissioner
DELAWARE	302/577-2885	benefits for volunteer based on their regular job wages
DISTRICT OF COLUMBIA	202/576-6265	NO VOLUNTEER FIRE FIGHTERS
FLORIDA	904/488-2713	benefits for volunteers based on
GEORGIA	404/656-3875	benefits for volunteers based on the GA average weekly wage for production workers in the manufacturing industry as determined by the GA Dept. of Labor
HAWAII	808/586-9151	benefits for volunteers based on regular job
IDAHO	208/334-2370	benefits for volunteers based on regular job(s)
ILLINOIS	312/814-6555	benefits for volunteers based on regular job
INDIANA	317/232-3808	volunteer partially covered: volunteer medical and disability costs paid; no lost time
IOWA	515/281-5934	benefits for volunteers based on regular job or amount = to 140% of statewide average for weekly wage, whichever is greater
KANSAS	913/296-4000	benefits for volunteers based on wage of comparable paid FF
KENTUCKY	502/564-5550	benefits for volunteers based on 2/3 of average weekly wage for regular full-time job up to $415.94/week
LOUISIANA	504/342-7555	coverage varies by jurisdiction
MAINE	207/287-7088	benefits for volunteers based on regular job

WORKERS' COMPENSATION LAWS BY STATE (Continued)

STATE/JURISDICTION	PHONE NUMBER	COMMENTS
MARYLAND	410/767-0900	benefits for volunteers based on maximum average weekly earnings allowable under statutes
MASSACHUSETTS	617/727-4900, ext. 357	statewide, volunteer fire fighters are not covered, unless individual city or town has its own policy to cover volunteers
MICHIGAN	517/373-3490	benefits for volunteers based on state average weekly wage
MINNESOTA	612/296-6107	benefits for volunteers based on wage for comparable paid FF in that geographic region
MISSISSIPPI	601/987-4200	volunteers are not specifically covered under Workers' Compensation, unless a contract were expressed or implied
MISSOURI	314/751-4231	benefits for volunteers based on wages of comparable paid fire fighter
MONTANA	899/336-8968 unit 2	coverage is voluntary; volunteer fire fighters can be covered under special policy endorsement, with compensation dependent on coverage selected (minimum $900/month; maximum $2280/month)
NEBRASKA	402/471-2568	benefits for volunteer based on regular job or, if not regularly employed by an outside person, 1 1/2 times the maximum compensation rate for total disability
NEVADA	702/687-5212	volunteers covered at a deemed wage of $2,000 per month if part of a regularly organized, recognized fire department
NEW HAMPSHIRE	603/271-3171	benefits for volunteers based on state average weekly wage
NEW JERSEY	609/292-2414	volunteers are covered with a benefit rate of the maximum applicable benefit set by law
NEW MEXICO	505/841-6000	coverage of volunteers is optional and varies between counties
NEW YORK	718/802-6666	benefits based on volunteer's regular job

WORKERS' COMPENSATION LAWS BY STATE (Continued)

STATE/JURISDICTION	PHONE NUMBER	COMMENTS
NORTH CAROLINA	919/733-4820	benefits based on average weekly wage for volunteer's principal employment
NORTH DAKOTA	701/328-3800	benefits based on wages of regular job
OHIO	614/466-2950	benefits based on wages of regular job
OKLAHOMA	405/557-7600	benefits based on wages of regular job
OREGON	503/387-7881	coverage of volunteers is not mandatory; municipalities can elect coverage
PENNSYLVANIA	717/783-5421	benefits for volunteers based on state average weekly wage
RHODE ISLAND	401/457-1800	volunteers covered under the State Board of Fire Fighter Relief @ $50/day
SOUTH CAROLINA	803/737-5700	coverage of volunteers is not mandatory and varies by county
SOUTH DAKOTA	605605/773-3681	benefits for volunteers based on maximum allowable under Workers' Compensation statutes
TENNESSEE	615/741-2395	coverage for volunteers is not mandatory statewide
TEXAS	512/448-7900	benefits for volunteers based on minimum amount off workers compensation allowable
UTAH	801/530-6800	benefits for volunteers based on average weekly wage
VERMONT	802/828-2286	benefits for volunteers based on wages of regular paid job
VIRGINIA	804/367-8600	benefits for volunteers based on maximum allowable under Workers' Compensation statutes
WASHINGTON	360/753-7318	disability benefits paid at $55/day, up to $1650/month, for 6 months; death benefit of $2,000; funeral benefit of $2,000; pension to surviving spouse of $825/month plus $70/child under age 18

E-5

WORKERS' COMPENSATION LAWS BY STATE (Continued)

STATE/JURISDICTION	PHONE NUMBER	COMMENTS
WEST VIRGINIA	304/926-5035	benefits for volunteers based on regular paid job (50% of daily wage for seven days per week) or, if volunteer has no paid job, then base benefits on minimum wage
WISCONSIN	608/266-1340	benefits for volunteers based on de maximum allowable under statutes
WYOMING	307/777-6932	benefits for volunteers equals 2/3 lost wages, subject to statewide average wage (maximum of $1837/month)

APPENDIX F

SUGGESTED INJURY REPORTING SYSTEM

INJURY REPORT

1.	RECORDER ID	FDID	STATION NO.	CASUALTY ID (SSN)		DATE		RECORD NO.
						Mo	Da	Yr

2.	NAME OF INJURED PERSON (LAST, FIRST, MI)	AGE	SEX	HOME TELEPHONE NO.	DAY OF WEEK	TIME OF DAY	JOB TITLE/RANK
			1) M 2) F				YEARS OF SERVICE

DAY OF WEEK: 1) Su 2) M 3) Tu 4) W 5) Th 6) F 7) Sa

3.	SHIFT	CONTINUOUS HOURS ON DUTY		NO. PRIOR RESPONSES/SHIFT	NO. PRIOR INJURIES/YR	TIME LOST ON JOB
		1) 1-12 2) 13-24 3) 25-36 4) 37-48 5) 49-60 6) 61-72 7) >72				1) none 2) 1Da 3) 2Da 4) 3-4 5) 5-7 6) 8 or more

SHIFT: 1) CURE TRANSFER/RESP. 2) ACAD/CHEMICAL 3) CHEMICAL RESP.

4.	GENERAL TYPE	GENERAL LOCATION ON BODY
	1) CURE TRAUMATIC 2) ACUTE CHEMICAL	1) SURFACE 2) INTERNAL 3) MIXED

5.	ACTIVITY AT TIME OF INJURY	WHERE INJURY OCCURRED	INJURY SEVERITY CODE (SEE DESCRIPTIONS BELOW)
	1) REST AT STATION 2) TRAIN AT STATION 3) TRAIN AT SCENE 4) EN TRANSIT TO SCENE 5) OTHER 6) ACADEMY TRAINING	1) AT STATION 2) AT SCENE 3) IN AMBULANCE TO 4) ENROUTE FROM 5) ROADWAY	1) MINOR 2) MODERATE 3) SEVERE 4) LIFE THREATENING 5) DOA 6) LATER DEATH

CAUSE ERGONOMICALLY CORRECT[?] ACTION TAKEN
1. YES 2. NO   1. STUDY   3. REVISED EQUIP.   2. REVISED JOB 4. OTHER

6.	INITIAL TREATMENT AT SCENE	ADVANCED TREATMENT	HOSPITAL (NAME)
	1) NONE 2) FIRST AID AND MEDICAL 3) FIRST AID ONLY 4) TRANSPORT	1) PHYSICIAN AT STATION 2) BY DEATH 3) ER/EQUIPMENT	PHYSICIAN (NAME)

ADVANCED TREATMENT: 1) PHYSICIAN AT STATION 2) EMT 3) EVALUATED AT HOSPITAL

7.	SPECIFIC GENERAL AREA OF BODY AFFECTED (SEE DETAIL ANATOMIC CODE	SIDE OF BODY (IF PERTINENT)
	1) MULTIPLE 2) HEAD 3) FACE 4) EYE/BOX 5) CERVICAL SPINE 6) THORACIC SPINE 7) LUMBAR/THORACO 8) LEG 9) FOOT 10) ARM 11) HAND 12) RESP/LUNGS 13) NECK/THROAT 14) HAZARDOUS GAS	1) RIGHT 2) LEFT

DETAILED CODE BY AREA TISSUE TYPE   WRITTEN DESCRIPTION   1. YES 2. NO

8.	DETAILED INJURY SITE CODE (SEE PAGE 2)		DETAILED INJURY TYPE CODE (SEE PAGE 3)

9.	ENVIRONMENTAL FACTORS	AMBIENT TEMPERATURE (F)	RELATIVE HUMIDITY (%)	PPE IN USE AT TIME
	1) CLEAR 2) CLOUDY 3) RAIN 4) SNOW/SLEET 5) WIND/DRY o 10MPH 6) HAZARDOUS GAS 7) POOR VISIBILITY 8) OTHER			1) SCBA/TURNOUT 2) RESPIRATOR 3) HELMET 4) EXPLOSION SUIT 9) OTHER

10.	SOURCE OF INJURY
	1) RADIANT HEAT 2) ENTRAPMENT 3) FALL 4) LIFTING 5) SMOKE 6) HAND TOOL 7) POWER TOOL 8) O. VEHICLE 9) OTHER VEHICLE 10) CHEMICAL EXPOSURE 11) EQUIPMENT FAILURE 12) FITNESS TRAINING 13) ATHLETIC CONTEST 14) PERSONAL EQUIPMENT 15) FALLING OBJECT 16) PULLING 17) STRUCTURE COLLAPSE 18) INFECTION 19) CLIMBING LADDER 20) OTHER

11.	EXPLANATIONS (NOTE SECTION REFERENCE OTHER?)

12.	DESCRIPTION OF INJURY

13.	NOTE (SEE ATTACHED)

INJURY SEVERITY CODE DESCRIPTIONS

1) MINOR–Patient is not in danger of death or permanent disability. Immediate medical care not necessary. This category includes most RSI, musculoskeletal injuries.

2) MODERATE–There is little danger of death or permanent disability. Quick medical care is advisable. This category includes injuries such as fractures or lacerations requiring sutures.

3) SEVERE–The situation is potentially life threatening. If condition remains uncontrolled. Immediate medical care is necessary, through body processes may still be functioning and vital signs are normal.

4) LIFE THREATENING–Death is imminent. Body process and vital signs are not normal. Immediate medical care is necessary. This category includes severe hemorrhaging and multiple internal injuries.

5) DOA–Dead upon arrival at scene. (or at hospital ER).

6) LATER DEATH–died subsequent to arrival (at scene or hospital ER).

DETAILED INJURY CODES BY AREA & ORGAN/TISSUE TYPE

INSTRUCTIONS: CODE BY → LEVEL #, SUBDIVISION #1 NUMBER, SUBDIVISION #2 NUMBER,
SIDE OF BODY IF PERTINENT → RIGHT = 1, LEFT = 2
WRITE DESCRIPTION & EXPLAIN ALL OTHERS LISTED.
MARK LOCATION OF INJURY (X) (IF PERTINENT) ON DRAWING

LEVEL #	SUBDIVISION #1	SUBDIVISION #2
1	1 CRANIUM	1. SKIN ONLY 8. NOSE
		2. BONE/CR. 9. URINARY
		3. MUSCLE 10. TEETH
	2 FACE	4. LIGAMENT/TENDON 11. TRACHEA
		12. ESOPHAGUS
		5. VASCULAR 13. CERVICAL VERT.
	3 NECK	6. NERVES/BRAIN 14. FACIAL BONE
		7. EYE 15. OTHER
2	1 SHOULDER	1. SKIN ONLY 15. LUMBAR VERT.
		2. BONE 16. ILIUM/ISCHIUM
		3. MUSCLE 17. SACRUM/COCCYX
		4. LIGAMENT/TENDON 18. RECTAL
	2 THORAX	5. VASCULAR 19. MULTIPLE
		6. NERVES 20. UNKNOWN
		7. HEART 21. OTHER
	3 ABDOMEN	8. LUNG
		9. LIVER/GB
		10. INTESTINE
	4 HIPS-PELVIS	11. KIDNEY
		12. GU.
		13. RIB
		14. THORACIC VERT.
5	1 THIGH	1. SKIN ONLY
		2. BONE
		3. MUSCLE
		4. LIGAMENT/TENDON
		5. VASCULAR
		6. FEMUR
	2 KNEE	7. PATELLA
		8. TIBIA
	3 SHANK	9. FIBULA
		10. CARTILAGE
		11. UNKNOWN
		12. OTHER
	4 ANKLE	
6	1 FOOT	1-5 AS ABOVE 7. METATARSAL
		6. TARSAL 8. TOE

SD #2	SD #1	LEVEL #
1. SKIN ONLY		
2. HUMERUS		
3. RADIUS		
4. ULNA	1 UP. ARM	
5. MUSCLE		
6. LIGAMENT/TENDON		3
7. VASCULAR	2 ELBOW	
8. NERVE	3 FORE-ARM	
9. UNKNOWN		
10. OTHER		
11. CARPAL	4 WRIST	
12. METACARPAL		
13. PHALANGES FINGER	1 HAND	4

DESCRIPTION: _____

OTHER (explain) _____

INJURY TYPE CODE

(FOR USE WITH INJURY REPORT, LINE 8)

1.	ABRASIONS
2.	ALLERGIC REACTION (AMBIENT AGENT)
3.	ASPHYXIATION (SMOKE/ POISON)
4.	BITE (HUMAN/ANIMALS
5,	BURN (CHEMICAL)
6.	BURN (THERMAL)
7.	CARDIOVASCULAR
8.	CEREBRAL VASCULAR ACCIDENT (CVA/STROKE)
9.	CONCUSSION
10.	CONTUSION/BRUISE (IMPACT INJURY]
11.	CUT/LACERATION.
12.	DEHYDRATION
13.	DERMATITIS (SKIN INFLAMMATION)
14.	DIGESTIVE (NAUSEA/VOMITING)
15.	DISLOCATION (JOINT)
16.	ELECTRIC SHOCK/BURN
17.	EXHAUSTION/FATIGUE
18.	FRACTURE (SIMPLE)
19.	FRACTURE COMPOUND
20.	FROSTBITE (FREEZING INJURY)
21.	GUNSHOT WOULD
22.	HEARING LOSS (TYMPANIC MEMBRANE)
23.	HEAT STRESS/HYPERTHERMIA
24.	HEMORRHALE????
25.	HYPOTHERMIA
28.	INFLAMMATION (SURFACE/UNKNOWN ETIOLOGY)
27.	INFLAMMATION (INTERNAL/UNKNOWN ETIOLOGY
28.	INTERNAL (UNKOWN)
29.	MULTIPLE (SPECIFY BY NO. OR LIST)
30.	POISON (INJESTED CHEMICAL)
31.	PUNCTURE (STAB WOUND)
32.	PYREXIA (FEVER/UNKNOWN ETIOLOGY)
33.	RESPIRATORY COLLAPSE (LIST CAUSE)
34.	RSI-REPETITIVE STRAIN INJURY/MUSCULOSKELEAL
35.	STRAIN/SPRAIN
36.	SPINAL (LIST TYPE & LOCATION)
37.	TRAUMATIC SHOCK
38.	VISION IMPARMENT,EYE
39.	OTHER (LIST)

APPENDIX G

EXAMPLE OF COST/BENEFIT ANALYSIS
(Riverside Fire Department, California)

RIVERSIDE FIRE DEPARTMENT

PHYSICAL FITNESS PROGRAM UPDATE
FEBRUARY 1989

In January 1987, with City Council authorization, the Fire Department implemented a mandatory physical fitness program. The program includes:

> Health evaluations
> Individual counseling
> Individual exercise prescription
> Daily on-duty exercise sessions
> Standards of performance

Program management, individual counseling, exercise prescriptions and minor injury consultation are provided by the contract agency, Community Orthopaedic Therapy and S.P.O.R.T. Clinic.

The City Council authorized the program for three years. To evaluate the benefits of the program, a study was implemented of costs for medical treatment and days off on workers compensation benefits. Statistics were compiled for the three years prior to program implementation. These statistics will be compared with the three years of the program.

Costs incurred for medical treatment showed a $36,000 reduction between the averages of the three pre-program years and the two program years. This reduction is equal to the cost paid to the contract agency for the three year program. Total cost for the program, including equipment, is $86,000.

Days lost in 1988 produced some interest statistics. Three injuries were responsible for 90% of the time off. One injury, a reoccurring heart condition, accounted for 60% of the days lost. Two back injuries account for 30% of the days lost. Sixty four injuries account for 10% of the days lost. The sixty four injuries included fifteen reported exposures to toxic substances or communicable diseases. The exposures are reportable injuries, but caused no time off.

The employees's heart condition originated in July 1984, before the physical fitness program began. It reoccurred in February 1988, and will cause retirement in February 1989,

The two back injuries accounted for 68% of the light duty days in 1988. They included herniated discs and bone spurs that were identified after increased physical activity; both employees completed corrective surgery. One employee has fully recovered and is predicted to reach maximum retirement age in ten years with no complications. The second employee is working light duty after corrective surgery with no predicted date for return to full duty.

1

Comparing two years of the program with the pm-program years showed the following:

* 23% reduction in medical costs

* 33% increase in total injuries

* 10% reduction in days off on workers compensation benefits

* 545% increase in light duty time

These statistics are encouraging when the unusual circumstances mentioned above are considered. The information developed from the two year participation in the program indicates that a physical fitness program can be cost effective.

The statistics compiled to date are shown on the attached graphs.

RIVERSIDE FIRE DEPARTMENT

WORKERS COMPENSATION INJURY STATISTICS

An analysis of information on injuries, time off on workers compensation benefits, and medical costs disclosed the following information:

YEAR	NUMBER OF INJURIES	NUMBER OF INJURIES WITH DAYS OFF	24 HOUR DAYS OFF	DAY OFF RANGE	DAY OFF AVERAGE PER INJURY	MEDICAL COSTS	HOURS OFF
1984	62	14	80	1-12	5.7	$200,882	1,920
1985	49	24	180	1-81	7.5	107,541	4,320
1986	52	16	190	1-63	11.9	165,810	4,560
1987	77	22	83	1-35	3.8	118,745	1,992
1988	67	12	184	1-106	15.3	124,150	4,416
1989 (8 mo.)	37	7	22	1-7	3.1	13,546	528

Each year listed, except 1989, had one or more major injuries which significantly influenced the average statistics for that year. The major injuries identified are:

YEAR	NATURE OF INJURY	NUMBER OF DAYS OFF	AMOUNT PAID IN MEDICAL COSTS
1984	HEART		$134,580
1985	STRESS	81	$37,446
1986	FALL	25	$ 26,303
	HEART	52	62,464
	BACK	63	34.982
1986 TOTAL:		140	$123,749
1987	CANCER	28	$ 66,419
1988	BACK	21	$ 22,108
	BACK	35	20,365
	HEART	106	(charged to 1984)
1988 TOTAL:		162	$ 42,473
1989	NO MAJOR INJURIES TO DATE (8 months)		

The major injuries identified in the program years 1987 and 1988 had origins in the pm-program years. The 1987 injury was the development of terminal cancer. The 1988 injuries were a recurring heart condition and two long term back injuries.

4

To provide a comparison between the three years prior to implementation of the Physical Fitness Program and the two years and eight months of the program, the actual statistics were adjusted by deleting the major injuries identified.

The following adjusted statistics projected below show dramatic improvement in the program years when compared with the pre-program years.

YEAR	NUMBER OF INJURIES WITH DAYS OFF	24 HOUR DAYS OFF	DAY OFF RANGE	DAY OFF AVERAGE PER INJURY	MEDICAL COSTS	HOURS OFF
1984	14	80	1-12	5.7	$66,302	1,920
1985	23	99	1-16	4.3	70,075	2,376
1986	13	50	1-17	3.9	42,061	1,200
1987	21	55	1-6	2.6	52,326	1,320
1988	9	22	1-8	2.4	81,677	528
1989*	7	22	1-7	3.1	13,546	528

Adjusted Statistic Averages:

1984, 85, 86	16.7	76.3	1-15	4.6	$59,479	1,832
1987, 88, 89*	13.9	37.1	1-7	3.0	55,261	889.9

* 1989: 8 months

Statistics developed show a positive benefit from implementation of the Physical Fitness Program.

DAYS LOST FROM INJURIES

PRE vs PROGRAM AVERAGES

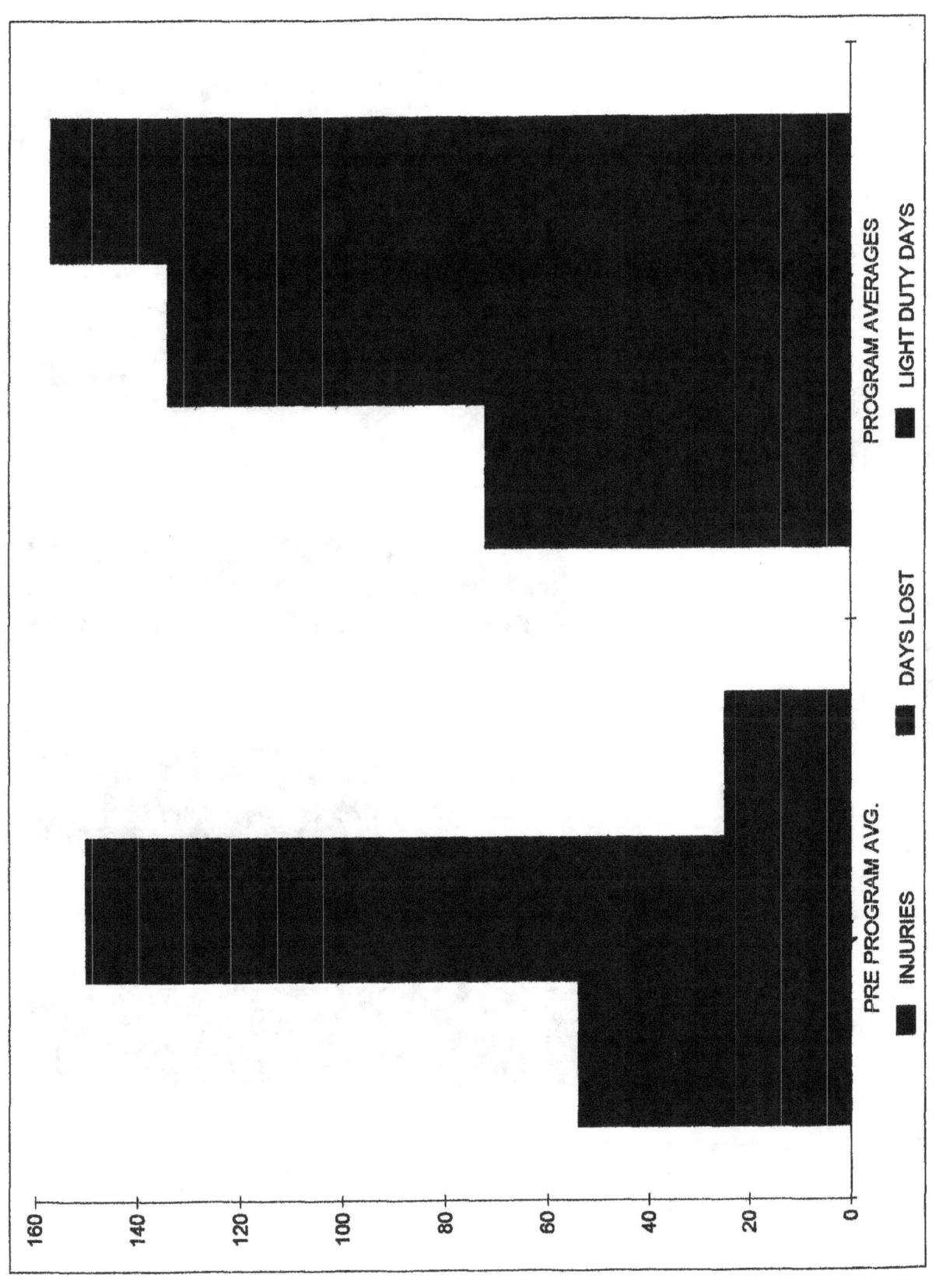

PRE PROGRAM AVG.   PROGRAM AVERAGES

■ INJURIES   ■ DAYS LOST   ■ LIGHT DUTY DAYS

84.85.86 VS 87,88

INJURY COSTS

MEDICAL COSTS PAID

PRE vs PROGRAM COSTS

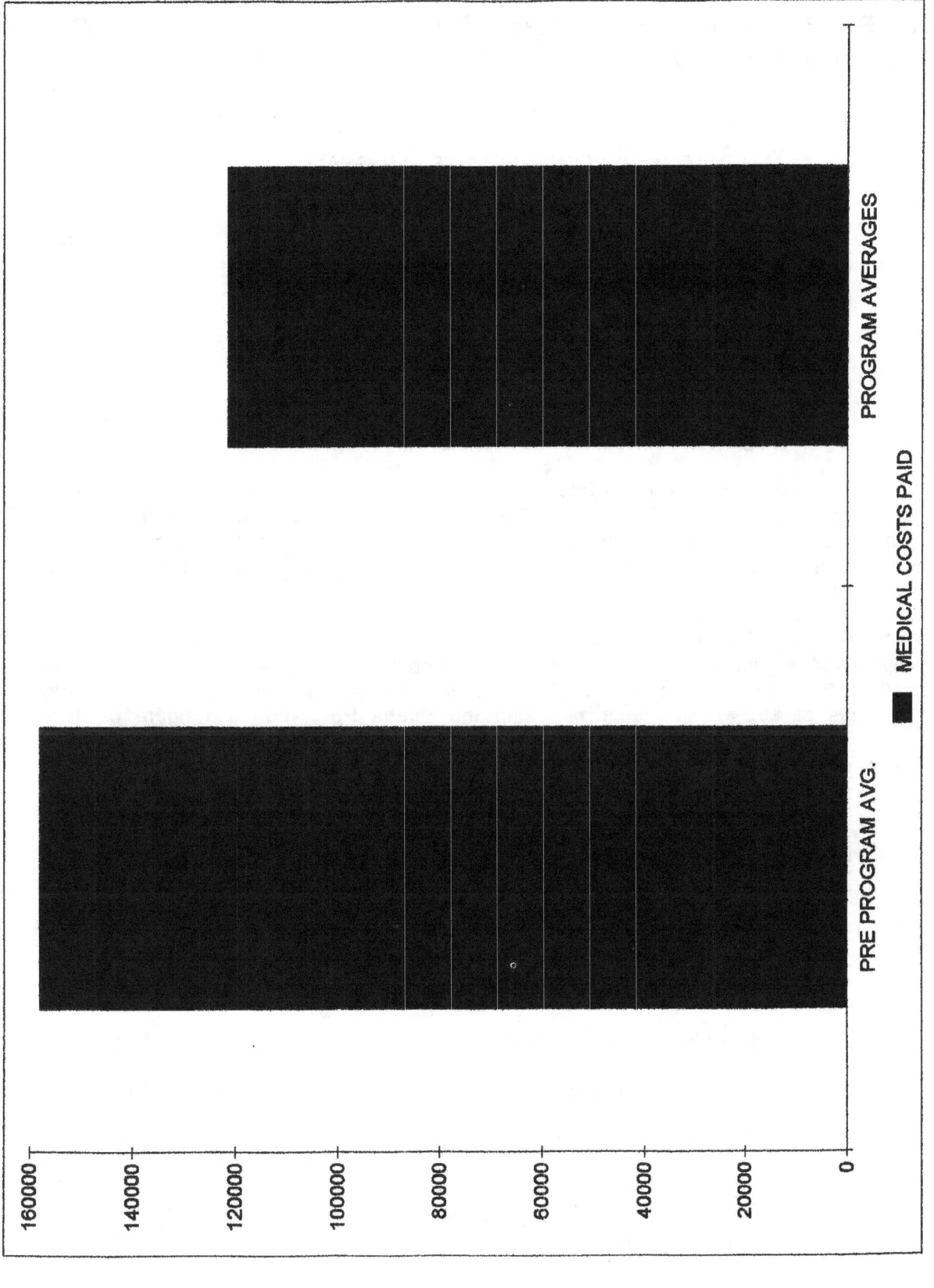

84,85,86 vs 87,88

CITY OF RIVERSIDE

CITY COUNCIL MEMORANDUM

HONORABLE MAYOR AND CITY COUNCIL

DATE: February 4, 1986

AGENDA ITEM:

SUBJECT: <u>MANDATORY PHYSICAL FITNESS PROGRAM</u>

The Fire Department Safety Committee has been actively pursuing the development of a comprehensive physical fitness program for four years. Consideration was given to:

- the need for a physical fitness program

- the strengths and weaknesses of current fire department fitness programs

- the expert assistance available to develop a program specifically addressing the needs of firefighters

Approximately ten fire department fitness programs within Southern California were studied. Only one, Buena Park, maintained statistics reflecting cost involved in their program. Buena Park showed a 37% reduction in time lost for industrial injuries in the 4.5 years following implementation of their program.

The City of Riverside Fire Department paid $735,000 for legal and medical and $233,364 additional wages for Workers Compensation costs from 1980 through 1985. The cost averaged $161,394 per year. Riverside can project a potential savings of $59,716 per year by using the Buena Park reduction figure of 37%.

The estimated costs for the program are as follows:

Year 1	Health evaluations for 176 personnel	$21,500
	Equipment & evaluations for 176 personnel	<u>41,000</u>
		$62,500
Year 2	Health evaluations for 71 personnel	8,700
Year 3	Health evaluations for 136 personnel	16,700
Year 4	Health evaluations for 111 personnel	<u>13,600</u>
	Total 4 Year Cost	$101,500

The benefit of implementing this program is reflected in the following projection:

Potential cost savings - 4 yrs x 59,716	$238,864
Estimated 4 Year cost	<u>-101,500</u>
Potential net savings	$137,364

This program is being negotiated with the Riverside Firemen's Benefit Association. It will become part of the Memorandum of Understanding if ratified by the membership. It will require mandatory participation by all Fire Department safety members.

The program is designed to produce the following:

- a health evaluation to identify potential health risks

- a fitness evaluation to determine individual levels of fitness

- a predetermined level of fitness for all personnel to maintain

- A planned program to assist personnel in progressing from current level of fitness to predetermined level of fitness

- a savings in workers compensation costs due to increased level of fitness of personnel

RECOMMENDATION

That the City Council authorize implementing the above described program for a two-year trial program.

PREPARED BY: Approved by,

D. M. Greene   Douglas G. Weiford
Deputy Chief, Fire Department   City Manager

DG/0054A/b

cc: City Attorney   Richard J. Bosted
    City Clerk   Fire Chief

CITY OF RIVERSIDE

CITY COUNCIL MEMORANDUM

HONORABLE MAYOR AND CITY COUNCIL   DATE: September 23, 1986

AGENDA ITEM:

SUBJECT: <u>FIRE DEPARTMENT - PHYSICAL FITNESS PROGRAM</u>

On February 4, 1986, the City Council authorized staff to pursue the implementation of a mandatory physical fitness program for all Fire Department Safety members. The proposed program has been negotiated with the Riverside Firemen's Benefit Association.

Request For Proposals were distributed to fourteen consultants to provide medical examinations and administer the program. Eight of the consultants responded with proposals. The four lowest bidders were interviewed by the Consultant Selection Committee composed of:

Doug Greene, Deputy Fire Chief
Engineer Gary Vanderhorst, President, Riverside Firemen's Benefit ASSOC.
Captain Paul Crawford, Past Chairman, Fire Department Safety Committee
Ms. Pat Rogers, Occupational Health Nurse

After reviewing the proposals and interviewing the consultants, the Committee recommends that Community Orthopaedic Therapy and S.P.O.R.T. Clinic be retained to provide the program.

The eight proposals submitted had costs ranging from a low of $36,385 to a high of $101,495 to administer the three-year program. Purchasing has quoted a price of $49,734 for the required equipment. The consultant recommended by the Selection Committee provided the lowest bid.

This program will be evaluated for cost effectiveness at the conclusion of the three-year period. The costs are as follows:

1986/87	Consultant Fees	$16,720
	Equipment	<u>49,734</u>
		$66,454
1987/88	Consultant Fees	6,745
1988/89	Consultant Fees	<u>$12,920</u>
	Total Program	$86,119

A summary of their proposal for the three-year program is attached for your review.

RECOMMENDATION

That the City Council:

1. Approve the retention of Riverside Orthopaedic/Therapy and S.P.O.R.T. Clinic to conduct the three-year mandatory physical fitness program for the Fire Department starting October 1, 1986,

2. Authorize the Legal Department to prepare a contract, and

3. Approve the transfer of $66,454 from the General Fund Contingency Reserve Account to the following accounts: $16,720 to 01-213-182, and $49,734 to 01-213-823 for fiscal year 1986/87.

PREPARED BY: Approved by,

R J Bosted   Douglas G. Weiford
Fire Chief   City Manager

Certified as to fund availability,   Concurs with,

H. E. Brewer   Lawrence E. Paulsen
Finance Director   Assistant City Manager

RJB/DMG/6152m/m
Attachment:
     Proposal

cc: City Attorney
    City Clerk

CITY OF RIVERSIDE

CITY COUNCIL MEMORANDUM

HONORABLE MAYOR AND CITY COUNCIL

DATE: February 6, 1990

AGENDA ITEM: 33

SUBJECT: FIRE DEPARTMENT - PHYSICAL FITNESS PROGRAM

On February 4, 1986, the City Council authorized staff to pursue the implementation of a mandatory physical fitness program for all Fire Department Safety members, and on September 23, 1986, the City Council approved the retention of Riverside Orthopaedic/Therapy and S.P.O.R.T. Clinic to conduct the program. It was the direction of Council that this program be evaluated after three years for its cost effectiveness. The original program costs were:

1986/87	Consultant Fees	$16,720
	Equipment	49,734
		$66,454
1987/88	Consultant Fees	6,745
1988/89	Consultant Fees	$12,920
	Total Program	$86,119

The three-year program is now complete and the following data is provided to show its cost effectiveness:

Three Year Period	Work Days Lost to Injuries	Lost Days Average	Cost for Lost Days Off	Employee Replacement Cost	Cost Saving Over Three Year Period
Pre-Program 1984	228.9	4.6	$81,448.40+	$122,232.60=	$203,721.00
During Program 1987-89	111.3	3.0	$39,622.80+	$59,439.20=	- 99,057.00
					$104,664.00

This program includes health evaluations, individual counseling, individual exercise prescriptions, daily on-duty exercise sessions and standards of performance. The Fire Department wishes to continue this program on a permanent basis due to its cost effectiveness and the benefit to our employees and our community. The Fire Department has $13,000.00 in the current budget for continuation of this program. No additional costs for equipment are anticipated.

RECOMMENDATION:

That the City Council:

1. Authorize the continued funding of the Physical Fitness Program as a permanent part of the Fire Department's operation.

2. Authorize the preparation and execution of a Request for Proposal soliciting interested firms to conduct the program.

PREPARED BY: Approved by,

Douglas M. Green   Douglas G. Weiford
Fire Chief   City Manager

DMG/2563e/a   Certified as to fund availability,

cc: City Attorney
    City Clerk   Barbara Steckel
                                          Finance Director

                                          Concurs with,

                                          Wendell Pryor
                                          Personnel Director

DAYS LOST FROM INJURIES

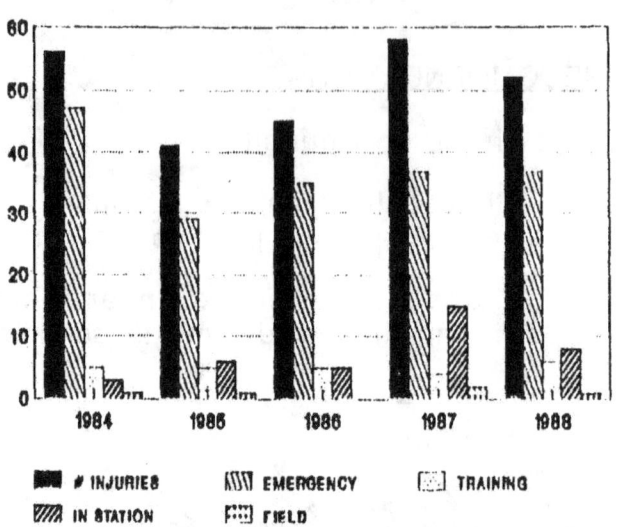

NON FITNESS INJURIES

Legend (left chart): INJURIES, DAYS LOST, LIGHT DUTY DAYS

Legend (right chart): # INJURIES, EMERGENCY, TRAINING, IN STATION, FIELD

FITNESS RELATED INJURIES

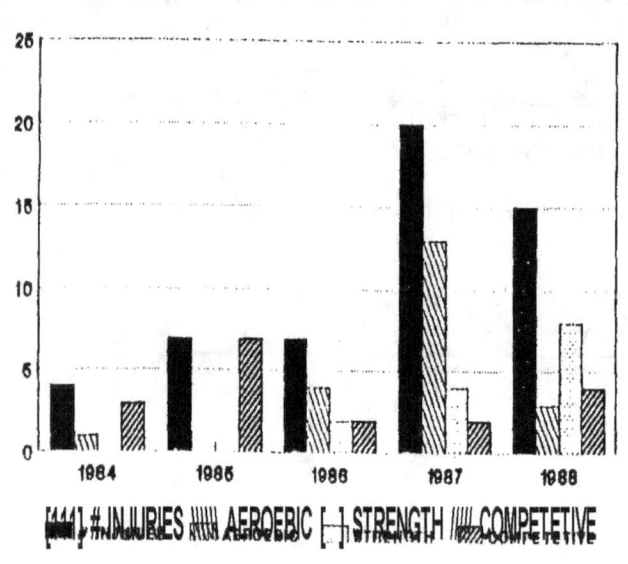

[###] # INJURIES AEROBIC [—] STRENGTH COMPETETIVE

PRE vs PROGRAM AVERAGES
84,85,86 vs 87,88

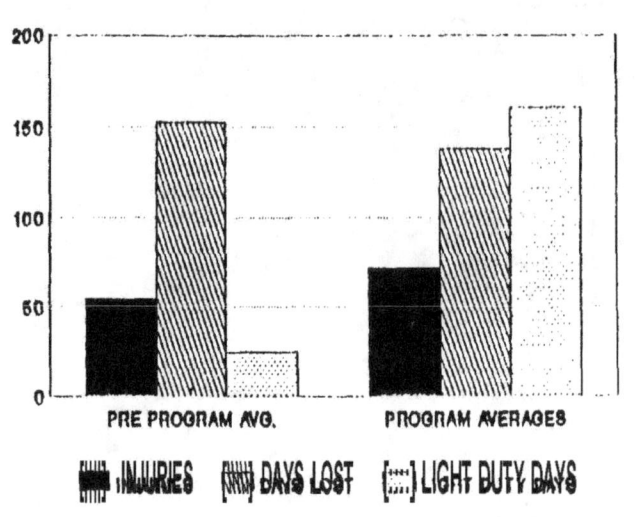

INJURIES DAYS LOST LIGHT DUTY DAYS

PRE vs PROGRAM AVERAGES
84,85,86 vs 87,88

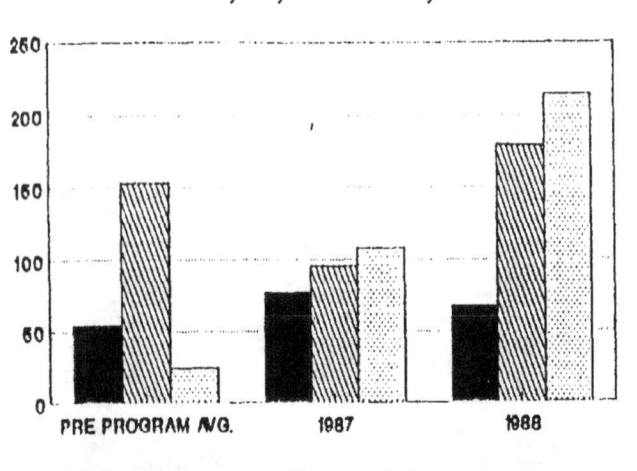

INJURIES DAYS LOST LIGHT DUTY DAYS

INJURIES BY AGE GROUP

INJURIES 20 TO 30 30 TO 40 40 TO 50 50 PLUS

PRE vs PROGRAM COSTS
84,85,86 vs 87,88

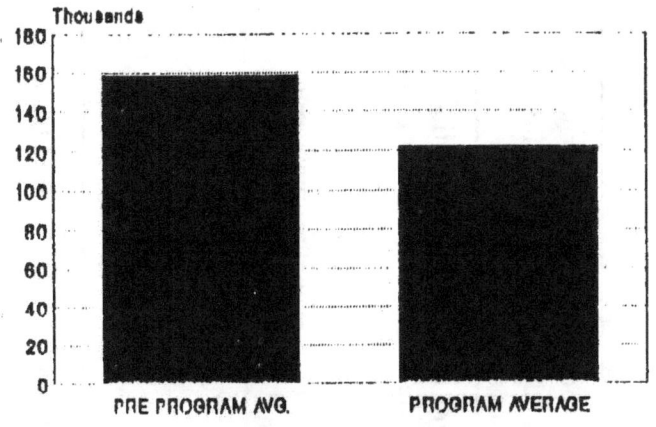

PRE vs PROGRAM COSTS
84,85,86 vs 87,88

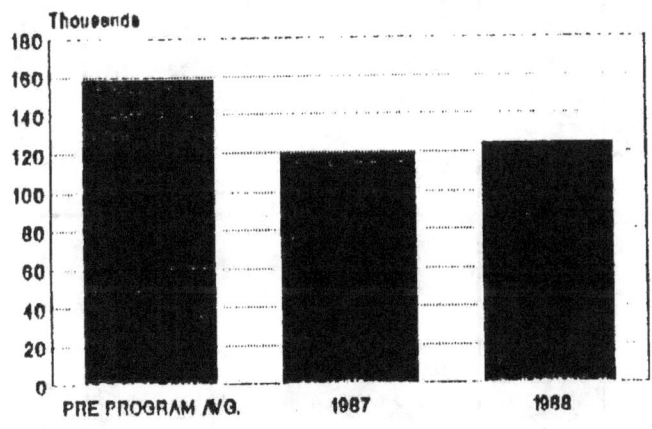

INJURY COSTS

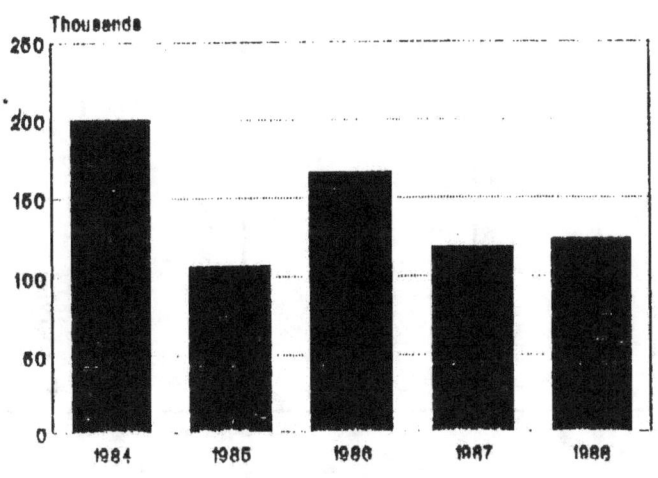

Table G-l. Representative Ergonomics Consulting Firms

Company	Address	Contact/Phone No.	Services
Advanced Ergonomics, Inc.	5550 LBJ Freeway Suite 350 Dallas, TX 75240	Kirby L. Dalton 214/239-3746, ext. 250 Joseph Selan, ext 204	analyze jobs, write ergonomics plan, train committee
Eidos Corp.	P. O. Box 30592 Lincoln, NE 68503	John Rudolph 402/466-1119	worksite and job analysis, ergonomics training and program implementation, rehabilitation management
Ergonomic Technologies Corp.	185 South Street Oyster Bay, NY 11711	Cynthia 1. Roth 516/922-2828	design and implement ergonomics program, including training; offer CURES™ (Cumulative Use Risk Evaluation System) and MESH™ (database for managing engineering, safety, and health programs)
Ergorisk Service, Inc.	1800 Walt Whitman Road Suite 600 Melville, NY 11747	Bob Salter 516/752-3550, ext 26	ergonomic consulting services
Ergostyle	9737 Aero Dr. Suite 100 San Diego, CA 92123	Kimberly Seymour 800/889-1777 619/999-4229 (voice mail)	ergonomic consulting and training

*This **is** not a comprehensive list of all consulting firms which provide ergonomic services. Appearance is this list does not constitute an endorsement of the subject firms b y the U. S. Fire Administration or Federal Emergency Management Agency.*

Table G-l. Representative Ergonomics Consulting Firms (Continued)

Company	Address	Contact/Phone No.	Services
GNA Corporate Health Professionals	950 Taylor Avenue Grand Haven, MI 49417	Brian Geers 616/846-1441	ergonomic consulting and training programs, and occupational rehabilitation
Humantech, Inc.	173 Parkland Plaza Suite D Ann Arbor, MI 48103	Scott Smith 313/663-6707	workstation audits and ergonomic survey s, corporate ergonomics program development and training, Risk Priority Management ("RPM") system
The Joyce Institute	1313 Plaza 600 Bldg. Seattle, WA 98101	Paul Simpson 800/645-6045	design and implement ergonomics program, train committee, ongoing management of programs
Woodward, Alpert, & Associates	515 N. Cabrillo Park Dr. Suite 101 Santa Ana, CA 92701	Joanne Alpert 714/565-3100	ergonomic worksite analysis, ergonomics team training, cumulative trauma injury prevention and back injury prevention programs
Workhealth Healthcare Management/Rehability Corp.	111 Westwood Pt. Suite 210 Brentwood, TN 37027	Ron Kerr 405/340-1855	job-site analysis, injury prevention and ergonomics programs, injury and rehabilitation management, ADA consulting

This is not a comprehensive list of all consulting firms which provide ergonomic services. Appearance is this list does not constitute an endorsement of the subject f irms by the U. S. F ire Administration or Federal Emergency M anagement Agency.